The Narrow Edge

The Narrow Edge

A TINY BIRD,

AN ANCIENT CRAB &

AN EPIC JOURNEY

Deborah Cramer

Yale UNIVERSITY PRESS/NEW HAVEN & LONDON

Published with assistance from the foundation established in memory of Philip Hamilton McMillan of the Class of 1894, Yale College.

Yale University Press books may be purchased in quantity for educational, business, or promotional use. For information, please e-mail sales.press@yale.edu (U.S. office) or sales@yaleup.co.uk (U.K. office).

Maps by Bill Nelson. Illustrations by Michael DiGiorgio.

Printed in the United States of America.

Library of Congress Cataloging-in-Publication Data
Cramer, Deborah.
The narrow edge : a tiny bird, an ancient crab, and an epic journey / Deborah Cramer.
pages   cm
Includes bibliographical references and index.
ISBN 978-0-300-18519-5 (alk. paper)
1. Red knot—Migration—Delaware Bay (Del. and N.J.) 2. Shore birds—Migration—Delaware Bay (Del. and N.J.) 3. Migratory birds—Delaware Bay (Del. and N.J.) I. Title.
QL696.C48C48 2015
598.156'80916346—dc23   2014040788

A catalogue record for this book is available from the British Library.

This paper meets the requirements of ANSI/NISO Z39.48-1992 (Permanence of Paper).

10 9 8 7 6 5 4 3 2 1

Contents

The Narrow Edge

Beginnings

One warm May night, around midnight, I drove out to an empty beach on Delaware Bay. The summerhouses nearby were dark and empty, the only light the full moon shining on the bay, and the only sound the waves gently lapping against the sand. Just before high tide, horseshoe crabs began emerging from the water. Their shells, some as large as dinner plates, were dark and scuffed. These prehistoric animals, emissaries from the deep sea, were coming in to lay their eggs in the sand. I'd never seen anything quite like this. I used to go down to the edge of the creek near my home in Gloucester, Massachusetts, to look for spawning horseshoe crabs, their unfailing arrival sign that a hard winter was turning to spring. There were never very many; at most I'd find six or eight. Delaware Bay is home to the world's greatest concentration of horseshoe crabs. On this beach, they came by the thousands, gliding effortlessly through the water, then burrowing in the sand. When the tide turned, they surfaced, slid into the waves, and disappeared. If I'd been at the beach an hour earlier or an hour later, I'd have missed them.

The next day more wildlife amassed on Delaware Bay beaches—thousands upon thousands of migrating shorebirds, an avian Serengeti, one of the greatest concentrations of shorebirds on the eastern seaboard of the United States. The birds remained in the bay for only a few weeks:

for many years, ornithologists didn't seem to know they were passing through. They'd come for the horseshoe crab eggs in flocks so thick I couldn't see the sand. Among them were a few thousand russet-colored sandpipers, red knots. They raced along the shore, frantically grabbing scattered horseshoe crab eggs. Where had the knots come from that they were so desperately hungry? And how could a diet of tiny eggs, each the size of a pinhead, take them where they were going? They wasted no time: they'd flown more than 7,500 miles to get here, and in two weeks, they'd be flying 2,000 more.

And that was only half their journey. Each year knots fly from one end of the Earth to the other and back. Consumed by curiosity, I followed them to learn what it takes to go such great distances, where they chose to stop along the way and why, and what was so special about those horseshoe crab eggs. This book is the story of that journey. I begin where many red knots live during the northern winter, a virtually inaccessible beach on the Strait of Magellan. When they begin flying north, I move with them, traveling to a crowded resort in Argentina, a saltwater lagoon in Texas, a hunting preserve in South Carolina. To see where knots build their nests in summer, I go to a lonely camp on Southampton Island in the Arctic's Foxe Basin, home to large numbers of hungry polar bears. When the breeding season ends and knots begin their long return to South America, I see them off, from the boggy edge of Canada's James Bay, the foggy Mingan Islands, a low-lying Cape Cod beach whose nearby waters are increasingly visited by great white sharks, and finally, the bay behind my home.

The journey is not easy. I accompany dedicated biologists and birders, tracking birds by foot, walking 10 or 12 miles every day through ice and snow. We sit for hours in the pouring rain, counting shorebirds. We hide on windy beaches hoping to catch them in nets. The knots are elusive. Fueled on fat, warmed by feathers, they can go anywhere, no matter how remote. We fly, too, watching for them from helicopters; listening for them in a small propeller plane equipped with a radio receiver; and following them onto the tundra with the help of bush pilots for whom a narrow strip of icy gravel constitutes a runway. We travel by boat, train, komatik, SUV, and ATV, on rides that range from exhilarating to hair-raising. I learn to load and fire, reasonably accurately, a 12-gauge shotgun, and find to my surprise on the next stopover that I miss it.

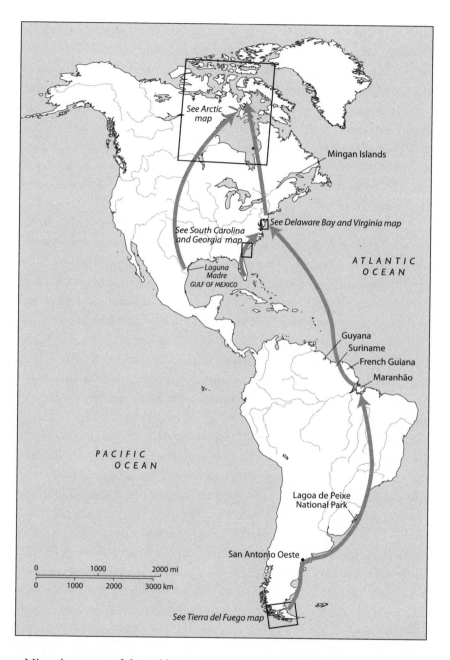

Migration route of the red knot, *Calidris canutus rufa* (map by Bill Nelson; source: U.S. Fish and Wildlife Service).

The knots seem at home in hurricane winds that ground us and in bug-infested, alligator-ridden swamps. I live in a mosquito-filled marsh but on this trip am subjected to the worst concentrations of biting insects I have ever seen. The birds forage for all their meals. Eating tiny clams and horseshoe crab eggs, they double their weight before each major flight. I taste their food, supplementing it with wild game, gourmet meals, dried crackers, and peanut butter—and lose weight. Slogging through isolated, remote areas looking for birds, I have a compass, GPS, and radio to keep track of myself. The birds have—what? By the end of this journey I am more in awe than when I began.

The route isn't quite what I thought it'd be. A few scraps of beach crowded with laughing gulls and shorebirds are a world-renowned hotspot for avian flu. One researcher there is funded by the Department of Homeland Security. In another state, I spend a morning not on a beach, but in a courtroom. Detouring off the well-marked path, I explore less recognized twists and turns that prove important, accompanying scientists as they uncover two previously unrecognized winter homes of young knots. Their work comes at a critical time. The U.S. Fish and Wildlife Service has listed the red knot, *Calidris canutus rufa*, as threatened under the Endangered Species Act; it is likely to become in danger of extinction in the foreseeable future. Along the route, I see why. Horseshoe crabs, I learn, matter as much to our own well-being as they do to shorebirds. I follow horseshoe crabs to a gleaming oyster bank in South Carolina, to a biomedical company in Charleston, and then to Massachusetts General Hospital to find out how and why my life depends on an animal that comes ashore but once a year.

The red knot whose migration I follow, *Calidris canutus rufa*, is one of six lineages of red knots worldwide. *Rufa*, the youngest of all the knots, flies the greatest distances. The birds have many homes, each a critical way station, a rung on a ladder between Tierra del Fuego and the Arctic. If only a few footholds break, the entire journey is compromised. Some are already broken. Some are being repaired, with hopeful result. Others are in danger of breaking. The story and struggle of *rufa* red knots is the story and struggle of all knots, and of millions of shorebirds. What would it mean if we lost them?

Migrating shorebirds speak to us. In the long arc of their journeys, in the soft, lilting calls of black-bellied plovers across a vast mudflat, in the hurried dash of sandpipers along the shore, they tell us of our world—what is, what is becoming, and what could be. In the quiet solitude to be found in the company of birds on a marshy island swept by an incoming tide, on a moonlit beach, in the cold, clear light of the Arctic summer—wherever we are relieved of the press of our busy lives—we can hear them and consider who we are and who we wish to be. Along a meridian that runs the length of the globe, I watch knots in their many homes, seeing firsthand how they live from day to day and how, as increasingly large numbers of people inhabit the narrow edge between land and sea, our lives are intertwined with theirs. Along the way, the story upends my ideas about guns, hunters, and hunting; about how humans and wildlife can share an increasingly crowded and redesigned shore; and, at a time when the edge between the human and natural world has dissolved, about what being wild now means.

I have always loved the many and exquisite ways science illuminates our world. With beauty and clarity, it offers insight into the lives of shore-birds and horseshoe crabs, and into our evolving shoreline. Science can suggest a direction, but science alone cannot repair our torn world. We make our choices from another plane. Following knots, I meet many dedi-cated people who year after year, season after season, are looking out for knots, striving to keep their homes at the edge of the sea safe and intact: scientists, birders, and people who love birds but don't call themselves birders; high school students, graduate students, rising biology stars, and those who've given their time to shorebirds for 30 or 40 years and continue even when they theoretically have retired. They work long days along a flyway that encompasses at least 12 countries, where people speak at least five languages, where the knot is known by many names, and where in one place, it has no name. They share a common dream, to restore to abun-dance a bird whose numbers have precipitously declined. Their service, rooted in science, springs from and is held by love.

In Delaware Bay I hold in my hand a knot that's flown the length of the Earth not once, but many times. This tiny bird has an unerring instinct to locate, over miles of coastline, individual beaches with the most plentiful food. It has developed astonishing ways to undertake, again and again, exhausting nonstop flights. It can bring forth a new generation in the harsh Arctic summer. Our human politics may vary, our needs and desires may conflict, and our values may differ, but this bird unites us along the shore of two entire continents, following a route that doesn't recognize our boundaries. I release the red knot, watching it take flight, praying it may continue to find shelter and refuge along its way, season after season and year after year. Hard questions lie before us, questions of how or whether humans and wildlife will share our increasingly fragile shore. Traveling through almost 120 degrees of latitude from the bottom of one continent to the top of another looking at those questions, I began to feel that, in the words of the Persian poet Hafiz, "All the hemispheres in existence lie beside an equator in your heart" where, perhaps, it is possible to see what is before us, hidden in plain sight.

One

THE "UTTERMOST PART OF THE EARTH"
Tierra del Fuego

Scientists Carmen Espoz and Ricardo Matus and their team load the ATV along with extra tanks of gasoline onto their truck. I follow them out of town. The newly paved highway quickly yields to a gravel road that becomes increasingly narrow, rutted, and dusty. I am thankful that before coming to Chile, I'd heeded last-minute advice to upgrade the rental car to an all-wheel drive with new tires. After 30 minutes on the road, we take an unmarked drive over small rolling hills and around a pond where flamingoes wade. A gaucho mends a fence while his horse, saddled in thick white sheepskin, grazes in the dry grass. In the distance a house belonging to Boris Cvitanic, owner of the estancia, nestles against a hill, amid the only trees to be seen. Cvitanic drives out to meet us. We manage to turn our vehicles in the tiny lane without digging up adjoining pasture. We then follow Cvitanic back onto the main road, down a sharp incline, and onto another unmarked lane, this one blocked by a fence. Cvitanic is a handsome man with a kind and gentle face. His jaunty beret, sweater, and sports jacket are all of fine wool. He chats with Espoz and Matus while he unlocks a giant padlock, and then gestures over the hill.

We push open the rickety gate and resume driving on the faint track, lurching over sharp rocks and through deep ruts. Grazing sheep dally in the road. We wait. The track winds around more low hills for a few miles and then straightens out along another fence. At a spot seemingly no different from anywhere else, we stop, unload the ATV, turn aside the fencing, and go through, beginning a hair-raising high-speed ride across one of the world's widest tidal flats. I can't tell where we are going or how we'll know when we've arrived. When Matus finally slows, we are far from shore. He leaves me on the mud and roars off to pick up the others. The empty mudflat reaches all the way to the horizon: the incoming tide is nowhere to be seen. As would happen so many times during this trip, and during the year that follows, I find myself in a remote place with landmarks I can't read, my companions people I barely know.

Time after time they would welcome me, these strangers working to protect the red knot—a small sandpiper about the size of a robin and weighing about as much as a coffee cup. The *red* in red knot is obvious enough: when they molt into breeding plumage their breast feathers turn rusty red. The origin and meaning of *knot* is more curious and obscure.

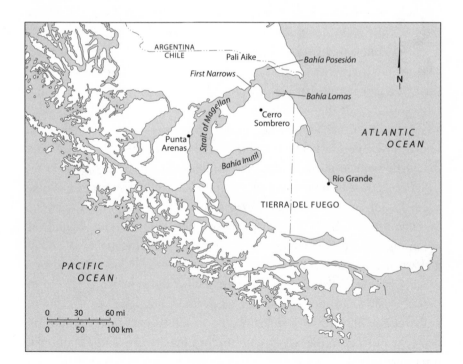

Tierra del Fuego (map by Bill Nelson).

Elizabethan poet Michael Drayton linked the knot to the eleventh-century Viking king Canute, writing that the knot "was Canutus' bird of old." Popular English dictionaries of the seventeenth and eighteenth centuries bolstered this idea, one describing knots as "a small delicious sort of small Fowl well known in some Parts of England, and so call'd from Canutus the Danish king, by whom they were highly esteemed."

Other accounts loosely incorporate the tendency of feeding knots to follow the tide. These tell of King Canute placing his throne at the edge of the sea and ordering the tide to ebb, disproving his acolytes' belief in his omnipotence. The *Oxford English Dictionary* finds these mythic connections "without historical or even traditional basis" and declares the origin of the bird's name unknown. The Chileans call the knot *playero ártico*, Arctic shorebird. Of all the names I would hear, this one is the most evocative.

For five months, red knots live on the beaches of Tierra del Fuego, feeding in the long sunlit days of the Southern Hemisphere's summer. As autumn approaches, they begin a long journey, 9,500 miles north to their breeding grounds in the Arctic, one of the longer avian migrations on record. They make the lengthy trip and its return every year. I have come to see them on this remote beach where they spend so much of their year and follow them along the edge of the sea from one end of the Earth to the other.

I am waiting on a stretch of isolated beach on Bahía Lomas, a broad bay at the Atlantic entrance to the Strait of Magellan, along the coast of Tierra del Fuego, one of Chile's most sparsely populated provinces. Icy fjords, steep mountains, and a 200-mile-long glacier, the Southern Patagonian Ice Field, cut this region off from the rest of the country. Getting here requires travel by air, sea, or taking a long drive through Argentina. Lucas Bridges, born in 1874 and raised on Tierra del Fuego by missionary parents, titled his classic memoir about the island he loved and the native peoples he lived among *Uttermost Part of the Earth*. More than 125 years later, the description still fits.

On this January afternoon, and probably most afternoons and mornings and evenings as well, the beach and road are empty: no cars driving by, no planes overhead, no tankers or boats in sight. No sound except for the steadily blowing wind. A cinnamon-colored guanaco leaps the fence and runs along the beach. Nothing breaks the broad reach of sand, open sky, and vast mudflat, whose tide had slipped more than four miles from shore. Espoz, dean of the Faculty of Science from the Universidad Santo Tomás in Santiago, Matus, a naturalist who has participated in research in Bahía Lomas for more than 10 years, and Laura Tellez, a field assistant from Punta Arenas, are awaiting the return of the water and of the shorebirds who feed at its edge.

My eyes tear in the wind. I'm squinting to shut out the sting. Espoz and Matus are wearing goggles. Maybe I won't even see, let alone hear, the birds. As the tide flows in, they could be anywhere along these 43 miles of privately held beach. Cvitanic, whose estancia we crossed, had seen a flock yesterday around 4:00 p.m. Cvitanic was born in Chile, but his father, a Croatian, had traveled far to make a home here. In the 1930s he'd crossed the sea to visit fellow Croatians who'd settled in Chile, fell in love with Tierra del Fuego, and stayed. "There was work," Cvitanic told me, and, "it was *tranquillo*, quiet."

His father went into sheep ranching and purchased an estancia further inland, next to a river, near what is now the international road to Argentina. Cvitanic took it over, sold it to his brother, and in 1994 bought a second estancia on Bahía Lomas. On 173,000 acres, he raises 5,000 sheep for meat and wool. Cvitanic remembers when he first saw the knots. "They'd come, suddenly, in the summer—enormous flocks, big clouds of birds—and then come winter, they'd disappear. They're not easy to find. I can't see them from the house, and the beach is so long. I'd see them come and go, and never realized they were coming from so far, never realized their special significance until the investigators asked permission to cross my land."

Northern Hemisphere ornithologists, on the other hand, knew where the birds came from but for years only vaguely understood where they might be going. Knots eluded naturalists who'd come here for the express purpose of studying and cataloging wildlife. Between 1831 and 1836, Charles Darwin served as naturalist to Captain FitzRoy of HMS *Beagle,* who'd been commissioned to survey waters along the South American coast. Darwin's extensive record of birds he observed and caught on Tierra del Fuego and Patagonia mentions plovers and other sandpipers, but not, from what I could find, knots. Between 1886 and 1889, Robert Oliver Cunningham, naturalist on HMS *Nassau,* commissioned by the British Admiralty to survey the Strait of Magellan, observed, shot, or attempted to shoot many birds, but red knots do not appear to be among them.

In 1904, Captain Richard Crawshay lived on Tierra del Fuego observing and collecting birds for the British Museum. He worked from estancia Caleta Josefina, on the west coast of the island, and estancia San Sebastian, on the east coast, the first sheep farm established by La Sociedad Explotadora de Tierra del Fuego (SETF). Owned and operated by the region's most powerful family, SETF would become the nation's largest and wealthiest ranching enterprise, eventually comprising 7 million acres. Like Cunningham, Crawshay shot many beach birds and wrote about them. Among other birds, Crawshay catalogued three species of plover, two species of oystercatcher, godwits, and white-rumped sandpipers—but no red knots. In all likelihood they were there, but despite careful observation, he missed them. He wouldn't be the only one.

Early twentieth-century records mentioned knots but were frustratingly vague. Robert Ridgway, curator of birds for the Smithsonian, and Alexander Wetmore, the Smithsonian's assistant secretary, traveled in South America to study birds. Both noted that the red knot's range extends as far south as Tierra del Fuego. Wetmore concludes that except in the vicinity of Buenos Aires, "the winter range of the knot has been little investigated, so that not much is known of the occurrence of this species." A. W. Johnson's 1965 *The Birds of Chile* says the knot is "among the rarest of visiting waders," and Rodolphe Meyer de Schauensee, in his 1966 *Birds of South America,* adds additional countries—Brazil and Uruguay—but doesn't specify where along this long shoreline the birds may be found. North America hosts 52 species of breeding shorebirds. When summer daylight begins to fade, more than half fly to South America. Not knowing their routes and destinations, scientists would be stymied if birds failed to return or if their numbers diminished. They felt called to unravel the mystery.

In November 1979, with support from the World Wildlife Fund, Guy Morrison from the Canadian Wildlife Service and Brian Harrington from the former Manomet Bird Observatory, now the Manomet Center for Conservation Sciences, set out to fill this nearly empty slate, undertaking a long drive from Buenos Aires down the east coast of South America. Prices were soaring: a two-seater Citroën was all they could afford. It caused trouble immediately. They hadn't left Buenos Aires when they heard a loud bang. Gunshot, Morrison thought. Everything turned gray. Morrison, whose sense of humor tends dry, briefly wondered whether he'd died. The latches on the front hood had snapped, causing the hood, which Harrington politely describes as being made of tin foil, to flip and wrap around the windshield. Their visibility completely obstructed, they were still on the road, still driving, Harrington at the wheel. Morrison, leaning out the window, frantically guided him away from oncoming trucks.

They pulled over, bent back the hood, tied it down, and drove on to Punta Rasa, a muddy tidal flat where the Rio de la Plata empties into the Atlantic. They'd been told there'd be thousands of knots. Six stops at Punta Rasa yielded 10, not an auspicious start. They were undeterred: a new, tantalizing paper by Belgian ornithologists Pierre Devillers and J. Terschuren described large numbers of knots at Rio Grande, Tierra del Fuego, 2,000 miles from Buenos Aires.

Their hopes high, they continued on, a blank map, knotwise, before them. After 1,000 miles and 15 stops, they'd seen only 20 red knots. Promising sites, like the Península Valdés, disappointed. Today, this promontory jutting out into the Atlantic and its 250 miles of coastline is a UNESCO World Heritage Site. Each year 80,000 people come to see an equal number of nesting penguins, breeding (and endangered) right whales, and colonies of elephant seals and sea lions, but in 1979, when access was more limited, they couldn't even see the beach for most of the drive. Further, Harrington told me, "The birds clump tightly together, so it's hit or miss." For the men, it was pretty much miss. Years later, when they knew more people there, they'd learn which tracks went down to the beach and whose land they could cross, but now, not wishing to trespass, they drove on.

Farther south, near Bahía Bustamante, the road followed a rise overlooking the distant sea before suddenly and fortunately dropping toward the coast. Two-thirds of the way down, a peregrine falcon soared. One of Earth's fastest animals, peregrines swoop down on their prey, usually a bird, at speeds of over 200 miles an hour. With eyesight two to three times as acute as ours, the falcon saw what Harrington and Morrison initially could not. With the peregrine sighting, their luck turned. On a broad, sandy tidal flat they found 400 knots, their first big flock. Walking slowly, they drew near 100 birds and saw 15 with oil-smeared chests and bellies. Though Bahía

Bustamante is quiet—60 years ago a man from Buenos Aires settled there to harvest seaweed and raise sheep—at the time, Comodoro Rivadavia—about 100 miles to the south—was a hub of Argentina's offshore oil production. Seeing so many oiled birds was disturbing. They drove on. Over the next 600 miles, they stopped 20 times and saw only 88 knots. Birding requires patience and perseverance. They had both. Spirits unflagging, they continued, buoyed by the expectation that around the next bend, at the next beach, they'd find something big. Looking back, Harrington admits that perhaps they were somewhat naïve.

The roads were awful, and the car was falling apart. They crossed the Strait of Magellan into Tierra del Fuego and arrived in Bahía San Sebastian. Crawshay had spent two months there, including October, when knots would be arriving, but didn't see any, while Harrington and Morrison whipped in for an hour to gaze at 250 knots and 640 Hudsonian godwits. Then they hurried on, and the next day reached Río Grande. Tourists visit Río Grande for world-class fly-fishing. Harrington and Morrison, hooked on birds, had no interest. After nine days and at least 2,000 miles of driving, their dream was fulfilled. Morrison recalls looking out the window from his hotel room and seeing a large flock of knots flying across the bay. Even though they'd visited 53 sites and found knots at only 11, the sighting at Río Grande was huge—over 5,000 knots feeding on the beach at low tide. They'd confirmed that large numbers of knots concentrate in a few areas, and that Río Grande was a major wintering site. They were satisfied. Unknowingly, however, they'd driven right by the mother lode.

After the inaugural road trip Harrington and he made through Argentina, Morrison, his appetite whetted, persuaded the Canadian government to fund a survey of the entire coast of South America, the South American Shorebird Atlas Project. This time, he and a colleague, Ken Ross, conducted aerial surveys, catching nooks and crannies and remote stretches along the shore Morrison had missed earlier. They flew almost the entire edge of the continent, looking anywhere they might find birds. They omitted Chile's southern tip, where the steep Andes abruptly rise from a labyrinth of dark fjords and densely forested islands. They skipped over Bahía Inutil—a bay on the western side of Tierra del Fuego explored in 1828 by Captain Phillip Parker King. King wrote that, entering the bay, he and his crew "flattered ourselves with the expectation of finding" an outlet leading to the Pacific. Instead, they discovered a dead end with "neither anchorage nor shelter," from which they "lost no time in retreating" and which they called Useless Bay. The name stuck. Bahía Inutil, lacking adequate shoreline for roosting birds, was also useless to Morrison and Ross.

Between 1982 and 1986, they flew 17,000 miles of coast, counting birds from helicopters or single-engine planes. To ensure consistency, they conducted the surveys together. Flying at high tide, 150 feet above the ground, at 100 to 150 miles per hour, they caught birds roosting at the water's edge. They were aloft anywhere between 8:00 in the morning and 4:00 in the afternoon. They taped their observations. The results were staggering. They found 2.9 million shorebirds, a far cry from the 6,200 Morrison and Harrington had seen on the road. The majority—2.5 million—were smaller sandpipers, "peeps," in Suriname and French Guiana. Knots prefer to winter farther south, in the "uttermost part of the earth." Almost every year, Morrison returns to count them.

He's scheduled a flight before I'm to meet Espoz and generously agrees to take me with him. We will leave from the northern side of the strait, from Compamento Posesión ENAP, a collection of houses, dormitories, and offices belonging to Empresa Nacional del Petróleo (ENAP), Chile's national oil company. The camp is set in a parched landscape of dry scrub where guanaco and flightless, ostrichlike rheas, locally called ñandú, roam in the dust. Foxes wander through the parking lot. An ashy-headed goose and her goslings graze in the grass. I join Morrison in the communal dining room for onces, a holdover from the English tea, served late in the afternoon. Over tuna sandwiches garnished with limp parsley and served with mayonnaise on soft white bread with the crusts cut off, Morrison reviews tomorrow's plans—to survey Bahía Lomas during the morning high tide. ENAP pilots, who ferry workers out to the rigs in one of the company's helicopters, will take us.

Morrison is on a tight schedule, having left one day for this particular flight. As we finish onces, the sky clouds over. At the end of dinner it begins to rain, and it pours all night. The storm pounds the metal roof of the guesthouse. Thunder and lightning roll across the wide open sky. Bad weather has its blessings: a few years earlier it brought Morrison to the ENAP camp. When he first began surveying the Strait of Magellan, he flew from Punta Arenas, 150 miles to the west. He'd hired a pilot, Victor Matus, to whom he'd advanced money for fuel. After they had circled the strait and surveyed Bahía Lomas, Morrison realized that the plane, blinking its lights and tipping its wings as it approached the runway, had no radio. On the ground, Morrison went home with Matus, whose son Ricardo, then 10 years old, still remembers his father bringing home "those big gringos." Ricardo now leads birding tours through Chile and conducts aerial surveys with Morrison.

One year, as Morrison completed the Bahía Lomas survey, a storm came in, its front too wide to detour around and its winds too strong to outrun. Clouds coalesced into a black mass broken by bolts of lightning. The plane hit a gust and dropped. The pilot called to the oil rigs for help and was offered an emergency landing at the ENAP camp if the plane could cross the water. ENAP now generously contributes helicopters, pilots, and logistic support for the shorebird survey. Leaving from ENAP, the trip is considerably shorter; Morrison will finish this run in about two hours. And helicopters have advantages. They travel more slowly, are more easily maneuvered, and their large windows offer a panoramic view.

The day Morrison and I are to fly dawns calm and clear. We meet the pilots at 8:00 a.m. and wait for the wind to push away a low cloud bank. The green helicopter is spiffy; the windows sparkle. Morrison is tall, but for years he's been folding himself into tiny cockpits with apparent ease and relish. He squeezes into his seat, dons his headphones, checks his tape recorder. We lift off and head west toward the ferry. The strait is calm, not calm enough to see whales and dolphins swimming below, but whitecaps are few. Brown cliffs rise from the edge of the mud. The tide washes in.

Twice we bank around a low gravel island. Knots sometimes roost there at high tide, but not today. At the ferry terminal we find some 1,500 knots on the gravel beaches. We turn south, crossing Primera Angostura, the First Narrows. The strait, 16 miles wide at the mouth, squeezes through a passage only two and a half miles wide at the First Narrows. Once across, we head east along Bahía Lomas. Flamingoes fly beneath us, bright pink against the mud. The helicopter flushes birds—small flocks, then larger ones. Morrison cranes his neck, whispers into his tape recorder. Reddish green algae coat the mud. Wisps of fog appear, and suddenly we're enveloped in a thick cloud. For eight minutes we fly suspended in the whiteness, seeing nothing, not even the beach. As we come out of the fog, the bay opens up. Narrow rivulets cut through the flats.

It was here, in January 1985, that Victor Matus turned his fixed-wing, Beech single-engine plane out along Bahía Lomas and Morrison discovered the largest concentration of knots wintering in Tierra del Fuego he'd seen yet. The 5,000 to 6,000 knots Harrington and he had seen from the ground at Río Grande paled before what he saw from the plane: 41,700 red knots and 10,520 Hudsonian godwits, many on mudflats along the estancia now owned by Cvitanic. "It was massive," Morrison recalls. "Not even the Chileans knew. It took our breath away." Bahía Lomas is the largest wintering site for *Calidris canutus rufa:* more than half the knots they saw in South America were concentrated on this one remote bay. The findings from this trip, large numbers of birds gathered in a few places, would suggest to Morrison a new way of thinking about shorebird protection.

For Morrison, counting birds from the air is an art, finely practiced and self-taught. "You have to start somewhere," he says, "so you begin small. You count 30 birds, get used to what that looks like. Then you build up to 50, to 100, then you group them, until you can see 1,000, 2,000, 5,000 birds. The tendency is to underestimate, so you have to be sure you don't compensate by overestimating." He's been honing these skills for years. From time to time, to check their accuracy, Morrison and Ross, who usually divided the survey between them by species, would count all birds, duplicating each other's work, and compare. Morrison has also tested his counting against computer simulations and not found it wanting.

The tide is sliding in. The bay is filling with water, broad stretches of flat yielding to the sea. Around 9:15, we approach Punta Catalina, where the strait empties into the Atlantic, and turn back, retracing our track. The flocks rise and fall: Hudsonian godwits, plump birds whose migration, only dimly known until recently, is as spectacular as the knots'; small sandpipers, too small to identify; groups of oystercatchers; and knots. They scatter as the helicopter approaches, then regroup. Mesmerized by great ribbons of birds, I don't even try to count.

Out on the beach, Espoz, Matus, and Tellez are studying the knot diet, identifying what is so nutritious, so plentiful it draws birds from thousands of miles away. Harrington and Morrison, whose own sustenance on the road consisted of bread, cheese, and wine, deduced knot sustenance from windrows of surf clams alongside feeding birds in Bahía Bustamante, and from feces packed with mussel shells in the roost at Río Grande. In Bahía Lomas, Espoz surveys mud. Beginning at high tide, she follows the ebbing water for more than a mile, taking samples of mud as she walks. Each time she stops, she takes mud from two depths—one the reach of red knots, the other the reach of godwits. Espoz is finding that knots prefer mussels and a small clam with a thin, almost transparent shell.

A puff of what looks like smoke rises in the distance: a flock too far away, too smudgy to identify. We wait. When Espoz comes to Bahía Lomas, she and her team work alone in the cold and the wind, surrounded by shorebirds. Like the birds, they come and go with the tide, beginning their surveys at dawn and returning late in the evening when the sun goes down. When they first began working here, they camped near the beach in the cold and blustery wind, miles away from fresh water. The wind, coming from the west, is constant. This far south, with little to break its momentum and intensity, it circles the globe unimpeded, often reaching hurricane strength. The scrub is low. Trees grow on a permanent slant. It's been blowing 80 miles per hour all week. Back in Punta Arenas, I'd been told repeatedly to park into the wind so the car doors don't blow off.

We may never know what originally brought knots to Bahía Lomas. Perhaps one long-ago day high winds swept them onto the beach, or perhaps they paused here, found the food plentiful, and stayed. Perhaps juvenile birds leaving the Arctic for their first journey south mistakenly confused Bahía Lomas with home. No one knows exactly when or under what circumstances the first birds arrived. Deborah Buehler's elegant reconstruction and analysis of the red knot's genetic family tree offers one history, which begins about 20,000 years ago at the height of the last Ice Age, when a small population of knots began evolving into the six different lineages known today. At that time, lobes of ice 2,300 feet thick filled Bahía Inutil and clogged the Strait of Magellan. Tierra del Fuego, including what would become Bahía Lomas, was tundra, nestled against the mainland. As the ice melted, 14,000 to 12,000 years ago, knots living in Siberia crossed the Bering land bridge into North America, perhaps following the same route humans had taken 1,000 or 2,000 years earlier. North American knots made their way south to wintering grounds in Florida and Mexico, possibly traveling, as the glaciers retreated, an ice-free corridor opening along the Continental Divide. Or perhaps, like the recent human arrivals in North America, they moved south along the ice-free Pacific coast.

Before red knots would find Patagonia and Tierra del Fuego and before the Strait of Magellan melted into a liquid waterway, humans may have made the long journey into southern South America. Bits of charcoal hint at their arrival in Bahía Inutil 13,000 years ago. They were living on the north side of the strait, near the volcanic crater of Pali Aike and the nearby shelter of Cueva Fell, 11,000 years ago. Pali Aike, hauntingly beautiful, rises above the plain along a rocky trail littered with black pumice. The trail climbs the back side of the crater, then plunges to its vertiginous edge. Baby ibis scream for food as adults shuttle back and forth from the rock cliff. Nearby, in a recess along a wall dotted with bright orange and red lichen, is the cave, quiet and sheltered from the wind.

Anthropologist Junius Bird unearthed human history here and at Fell's Cave in 1936 and 1937. During his long career at the American Museum of Natural History, Bird, like the knots, traveled the length of the Earth, working in both southern Chile and in the Arctic's Southampton Island, where knots nest. Bird spent two summers at Pali Aike and Fell's Cave, digging up bones, tools, and early weapons, and living on mutton, roasted and fried, homemade doughnuts, steamed chocolate pudding, and the occasional pot-roasted ibis.

At Pali Aike, he found cremated remains of humans. In Fell's Cave, he found ancient hearths with broken and burned bones (enough to fill a large garbage can) of ground sloth and native horse. Bird uncovered the history of a time, 11,000 years ago, when sloths roamed the South American steppes and humans slaughtered and ate them. The giant animals and recently arrived humans would not share the Earth for long. Somewhere between 11,000 and 7,000 years ago, 37 species of giant animals, each weighing more than a ton, disappeared from South America. Glyptodonts—animals that looked like armadillos but were sized like dinosaurs—giant ground sloths the size of elephants, and the smaller native horse became extinct.

At the time, the sea lay 125 miles farther east; coastal settlements containing history of this time now lie submerged. Argentine paleontologists find that what record there is tells of retreating glaciers, a warming climate, and dwindling grasslands. Large animals, increasingly stressed as their habitat shrank, succumbed to human hunters, perhaps in one or two millennia. The extinction of large animals at the end of the Ice Age was not the first, and would by no means be the last, case of humans pushing animals over the edge.

Red knots wintering in Bahía Lomas may never have flown over a giant sloth. The paths of the small birds and giant animals, one species arriving as the other departed, may never have crossed. They lived in different worlds. As the large animals disappeared, the ice continued melting. The sea rose, and 8,500 years ago, it inundated the continent. Atlantic and Pacific waters joined in what would become the Strait of Magellan, drowning the land bridge to Tierra del Fuego. The rising sea created one of the world's most spectacular submerged shorelines—a tattered coast of dark fjords shadowed by thousands of mountainous islands falling steeply into the water. Sea level has dropped 11.5 feet since then, but the coast is still ragged, by Bird's estimates 12,000 miles of coastline for 1,000 miles as the crow flies.

The receding sea exposed the wide mudflats that red knots love. Whether they arrived 5,000 or 500 years ago scientists can only speculate. When they came, Indians were using the wind-protected cave at Pali Aike, surviving on the dry plain by killing guanaco with small stone bolas, eating the meat, and making tools from the bones, thread from the sinews, and tents and clothing from the skins. They lived on the Patagonian and Fuegian steppes for 12,000 years. After the Spanish arrived, as archaeologists Mónica Salemme and Laura Miotti describe, "It was a slow but constant and brutally aggressive process that in 200–300 years . . . destroyed a hunter-gatherer life way more than 12,000 years old."

Cvitanic has steered us to the right place along the long beach at Bahía Lomas. After two or three hours, the tide approaches, washing in over the mud. With it come the birds. Hudsonian godwits and red knots wheel overhead, circling in flocks of 1,200, 2,000, 5,000. Each flock moves in unison, individual birds rising, turning as one in smooth, sinuous curves tuned to music I cannot hear and cues I cannot see. In an instant, a whole flock curves upward, the birds' white bellies gleaming in the sun. Then it levels off, brown against the gray sky, each bird evenly spaced, one never jostling another, never breaking the curve—a seemingly impossible precision and unity. We stand quietly and as the water rises, the birds land. I am surrounded by more than 1,000 birds. They have flown 9,500 miles to reach this beach. Some are juveniles. The adults are ashy gray, their breeding plumage faded. How they get here unaccompanied by their parents defies comprehension.

How migrating animals find their way is still one of science's big unanswered questions. Baby loggerhead sea turtles hatching in the sand along the coast of Florida and the Gulf of Mexico find their way to the water and then ride currents across the sea to the Mediterranean, where they will grow up. Their mothers departed long before their eggs hatched, but they passed down a set of directions for the baby turtles to follow. The famed sea-run brown trout sought by fly fishermen in Río Grande may navigate to the streams of their birth using a compass of iron-rich magnetite crystals housed in their noses. Homing pigeons disoriented by supersonic blasts from the Concorde may steer by low-frequency sound waves. Migrating birds navigate by the motion of the sun by day and the stars by night, and some, with compass needles embedded in their brains, follow curves and variations in Earth's magnetic field. Perhaps juvenile red knots and godwits traveling to Bahía Lomas are born with a map and instructions passed down from their parents, imprinted in their genes.

Ferdinand Magellan, like the young knots, sought a place he didn't know, finally finding it after considerable time and duress and with the assistance of many navigational aids. Knots leave Canada in August and begin arriving in Bahía Lomas by October. Magellan weighed anchor from the port of Seville on September 20, 1519, arriving in the strait that would bear his name more than a year later. He carried "23 charts . . . 7 astrolabes (one of brass), 21 wooden quadrants . . . 35 magnetized needles for the compasses, and 18 half-hour glasses." The birds depart from their last refueling stopover, possibly in Brazil, with only their fat to carry them through. Magellan loaded his ships with "wine, olive oil, vinegar, beans, lentils, garlic, flour, rice, cheese, honey, sugar, anchovies, sardines, salt cod, salt beef, salt pork," along with "live cattle and swine to slaughter en route" and boxes of *membrillo,* a quivery, not always that tasty quince jelly still sold in blocks in the markets in Argentina. His stores would not last. He frequently resupplied from the abundant marine life. At one point crew members slaughtering penguin and elephant seal were stranded as their ship was yet again blown out to sea. As they awaited the ship's return, they escaped a sure death huddling beneath the smelly carcasses. Knots weather the cold warmed by their fat and feathers.

Time and time again, Magellan was blown back by raging gales, barely managing 120 miles in three weeks. Knots travel the length of continents in that time. How they persevere we cannot say. Starving, hungry, buffeted by gales, and plagued by doubts, Magellan's crew lost faith. Concluding that the long-sought strait was an illusion and the mission doomed, the men mutinied. The captain, determined to find his passage to the Pacific, had their leaders decapitated, drawn and quartered, hanged, or marooned and left to die on the desolate shore.

Sailing into the strait on October 21, 1520, Magellan sent two ships to hazard the fast-running tides, dangerously narrow channels, and treacherous shallows and left two in Bahía Lomas. The birds would have been arriving about then. Perhaps his crew saw knots circling overhead, alighting on mud exposed by the receding tide, or perhaps these birds, not carrying as much meat as penguin or elephant seal, escaped their notice.

Magellan, at least, succeeded. Others were not so lucky. Simon de Alcazaba, sent by Spain in 1534 to explore the coast of Chile, sailed past Bahía Lomas and made it through the First Narrows before a gale forced his ship back 80 miles into the ocean. With water and supplies low, the cattle survived on wine and the crew on pickled penguin. After Alcazaba attempted one more foray through the First Narrows, the crew mutinied, and the ship returned to Spain. In 1557, Cortés Horjea tried from the western side, sailing through one icy fjord after another and into bay after bay that had no outlet. One dead end he named Seno Ultima Esperanza—Last Hope Sound. After two months of being tossed by williwaws—blasts of cold air plunging from the mountains into the fjords—his crew and he lost their anchors, ran out of food, and gave up. Four hundred years later, the bad weather had not abated. In February 1896 Joshua Slocum, sailing solo around the world, entered the strait, and after deftly navigating racing tides and fierce currents, was caught in a gale that "struck like a shot from a canon" and raged for 30 hours. His map locates the storm between Bahía Posesión and Bahía Lomas, the two beaches where I accompanied Morrison on his aerial survey.

Satellite transmitters, tracking larger shorebirds flying through the vortex of dangerous storms, begin to suggest how they manage brutal winds. In August 2011, amazed scientists tracked a whimbrel named Hope flying 90 miles per hour through a tropical storm. Later that month another, Chinquapin, sailed through Hurricane Irene's 115 mile-per-hour winds. In 2012, a third whimbrel, Pingo, detoured around Hurricane Isaac. Winds can fling birds far afield, creating a "windfall," literally, for birders accumulating life lists. At the same time birds may expend extra energy avoiding storms: it helps if they are well provisioned beforehand. Hope flew nine miles an hour for 27 hours before she caught the storm's tailwinds. Two red knots migrating south in the fall each flew an additional 600 miles to skirt high winds.

Magallanes, the area around the Strait of Magellan, hosts small sandpipers that fly great distances, penguins and ducks that can't fly at all, and one of the world's largest flying birds, the condor, so heavy it needs the wind to keep it aloft. When naturalist Cunningham surveyed the Strait of Magellan, he found condors perched on the cliffs of Bahía Lomas. Darwin observed them riding currents of air rising off sun-warmed rock, in languid, seemingly effortless flight, gliding for half an hour without flapping their broad wings. I wonder what he would have made of the long and slender wings of knots which, adapted to go great distances, keep the birds aloft for days at a time.

The tide is turning. It laps the edge of the grass and then gently glides away. Everyone looks up as a rare Magellanic plover whizzes by. The sun is setting. Realizing I may never return here, I absorb as much as I can of this harsh but soothing place where the tide flows uninterrupted, where time passes unhurried, and where the birds seem at peace and undisturbed. Tierra del Fuego may be their safest haven, yet in recent years, their numbers have been disconcertingly low.

Two

WHEN IS THE BEGINNING OF THE END?

Human history is already marked by the passing of birds whose earthly tenure we shortened, by birds returned from the brink of extinction by sheer human will and determination, and by birds that teeter on the edge. John James Audubon, on his way to Louisville in 1813, witnessed thick flocks of passenger pigeons flying overhead, flocks so big that three whole days elapsed before the birds passed by and the sky lightened. In three hours he saw more than 1 billion birds. Accounts of passenger pigeons in the woods of Massachusetts not more than 10 minutes from my home tell of millions and millions of birds, and of pine trees so loaded with nests "the Sunne never sees the ground." I never knew these birds, never saw one, let alone a flock or a tree whose branches creaked with the heaviness of nests. I am left with old histories of those who did and a forest filled with ghosts, emptied of lives unnecessarily lost. It's difficult to grasp, as it must have been for the hunters who killed the passenger pigeons and the westward-expanding nation that cleared their forest homes, that so many birds could be eradicated, yet by 1914 the passenger pigeon was gone.

For something as momentous as extinction, the end, when it comes, can pass with neither mourning nor remark. In 1967, in Newfoundland's coastal village of Port-aux-Choix, a bulldozer digging up sod in preparation for the construction of a theater and billiard hall uncovered an ancient cemetery belonging to maritime Archaic Indians who'd lived there 4,700 years ago. Alongside the human remains, archaeologists unearthed hunting tools and 30 species of dead birds. Three-quarters of the individual birds were great auks: large, flightless, deep-diving sea birds that once bred on rock islands off Newfoundland, Scotland, and Iceland. One site held 200 great auk beaks once attached to a cloak of auk skins, intimation that the great auk may have been revered in a way now lost to us.

The glossy black-and-white birds survived for thousands of years and then succumbed as we slaughtered them for food, bait, mattresses, and quilts. The last great auks were killed in 1844 on the island of Eldey off the coast of Iceland. One early June night, hunters set out from shore, arriving at the base of the island's rock cliff early the next morning. According to an account published in 1861, in "much less time than it takes to tell," three men went ashore, chased and easily captured two auks, left behind a broken egg, hurried back to the boat, strangled the birds, and then, as the wind rose, cast off. That was that. The great auks were going to a collector. The hunt was commissioned not for feathers or food but for skins. These mythic polar birds of a distant time and place once ranged as far south as Florida and, like passenger pigeons, alighted not far from my home on a barrier beach where their bones have been found in Indian middens buried in the dunes.

What of the ivory-billed woodpecker? This bird may disappear in my lifetime if it hasn't already. It lives or lived in old-growth forests of the South, stripping bark from dead and dying trees to chisel out the larvae of wood-boring beetles. The last confirmed sighting was more than 60 years ago in Louisiana. Since 2000, there've been reports of ivory-bills seen and heard in bottomland forests and swamps in Arkansas, Florida, and Louisiana, but the sightings are not definitive. The bird is gone from my newest birding guide. Other guides acknowledge that incontrovertible evidence is lacking, yet leave the entry, a hope rationally or irrationally held. Perhaps keeping the ivory-billed woodpecker in a guide is wishful thinking, denial that we snuffed out a life that can be no more, and our reluctance to face the enormity of actions irrevocable. I dream on, hoping we live in a time when we haven't uncovered everything there is to know and haven't seen everything there is to see, and that hidden in a cypress swamp somewhere a few birds, unknown to us, are holding out.

The spoon-billed sandpiper is on the edge of extinction. Sandpipers are notoriously difficult to identify, but this one's beak distinctly broadens at the tip, if not into a spoon shape, then into something resembling a spatula. Fewer than 200 breeding pairs of this critically endangered bird are left. Their fall has been swift, dropping by an order of magnitude since the 1970s. Spoon-billed sandpipers live in at least three places essential to their health and well-being: the Siberian tundra where they breed, the coasts of Bangladesh and Myanmar where they winter, and the tidal flats along the Yellow Sea where they refuel. Their homes are stressed. In Bangladesh and Myanmar, poverty-stricken villagers trapping egrets and herons with nets and potassium cyanide bait have also snared smaller plovers, curlews, and spoon-billed sandpipers. En route to the Arctic, spoon-billed sandpipers refuel on mudflats fringing the Yellow Sea. Mile after mile of seawall has "reclaimed" Yellow Sea tidal flats for intensive aquaculture and for factories and infrastructure to serve rapidly expanding Asian cities. Half the Yellow Sea's tidal flats have been destroyed, hurting not only spoon-billed sandpipers but other shorebirds as well—dowitchers, curlews, dunlin, and bartailed godwits—that breed in Alaska and winter in Asia and whose numbers are also declining.

A full-scale effort is under way to save the spoon-billed sandpiper. Between 80 and 90 percent of the hunters in Myanmar who have been offered fishing boats, livestock, or other means to secure food and income are taking fewer birds. Ornithologists retrieve spoon-billed sandpiper eggs from the wild, incubate them where they are safe from predators, and then release the fledglings to fly south with their siblings. To the relief and joy of the scientists, one fledgling set free on the tundra in 2012 migrated 5,000 miles to Asia, survived its first two winters, and then was sighted in May 2014 en route to Siberia, where it will breed for the first time. There, photographer Gerrit Vyn has filmed what most of us will never see: baby spoon-billed sandpipers emerging from a nest, wobbling off to peck for food in tundra grasses, and a protective father calling his chicks back before they get too cold.

For many shorebirds migrating north from Asia, the Yellow Sea is the last stop before the Arctic, the last place where thin, tired birds gain strength and energy for the long flight and breeding season ahead. Deteriorating tidal flats along the Yellow Sea jeopardize the lives of 2 million migrating shorebirds. Many are trying to save this fertile and essential resource before it is fully eliminated. The work tests whether 1.4 billion people living along the Yellow Sea will share its shore, and whether an extinction that had seemed inevitable might be delayed and possibly averted.

The fierce determination of wildlife advocates has safely turned birds from the path leading to extinction. At the turn of the nineteenth century, hunters shot at least 5 million ibis, heron, and snowy, reddish, and great egrets every year, taking their beautiful cascading plumage to adorn the hats of fashionable women. The nation's first Audubon societies, the American Ornithological Union, and legislation prohibiting the hunting of migratory birds were born from this excess. Aristocratic Boston socialite Harriet Lawrence Hemenway found the carnage appalling. Over tea with her cousin Minna B. Hall, these mothers of conservation, poring over the Boston Blue Book with its list of Boston's elite, enlisted 900 women of wealth and power to boycott feathered hats and formed the Massachusetts Audubon Society. The gorgeous, showy birds are still with us. Often, on an early autumn day, when the marsh by my home is turning a golden yellow and the air and water are still warm, I paddle by 20, 30, sometimes 50 or 60 or even 100 snowy egrets standing in the golden grass. Their absence now would leave a quieter, sadder landscape.

The bald eagle, peregrine falcon, and brown pelican were casualties of DDT. The birds, eating pesticide-contaminated prey, laid thin-shelled eggs that cracked under the weight of incubating birds. Publication of Rachel Carson's *Silent Spring* and a subsequent lawsuit by the newly formed Environmental Defense Fund ultimately resulted in a ban on DDT, setting the stage for the birds' dramatic recovery. Brown pelicans rebounded. I've often seen crowds of them along Florida beaches, perched on pilings and docks or gliding along the water, their pouches heavy with fish, an abundance once unimaginable. Closer to home, I've stood beneath old pine trees as a huge eagle perched above me, and I've gazed at adults and young circling over the river nearby. From the fishing pier in Gloucester I have watched a peregrine that lives on a ledge at the top of city hall. On my travels I see them swooping at red knots. It took perseverance and dedication to restore these endangered birds: peregrine falcons were on the U.S. Fish and Wildlife Service's (USFWS) endangered species list for 29 years, the brown pelican for 39, and the bald eagle for 40.

After 50 years, whooping cranes are still on the list. North America's tallest bird, they reach five feet in height. The one wild population, about 300 birds, breeds in Canada's Wood Buffalo National Park and winters along the Gulf Coast of Texas in the Aransas National Wildlife Refuge. When 23 cranes died there in the winter of 2008–9, the Aransas Project sued the Texas Commission on Environmental Quality, arguing that regulators were permitting industry and municipalities to take too much water from the San Antonio and Guadalupe rivers, leaving too little for the refuge and its inhabitants. Blue crabs and wolfberries, the cranes' principal food, couldn't thrive in the salty water. Cranes, lacking both freshwater to drink and their customary food, grew weak and died.

I sat in on the trial and listened to the arguments. One morning I went by boat into the creeks and marshes of the refuge to find whooping cranes. A few flew overhead—large white birds with black wing tips, long thin necks, and dark cheekbones. That this population had been rebuilt from only 15 birds seemed nothing short of miraculous. Two parents stood in the dry grass with their chick. The guide said that ponds where the cranes fed had dried up, that blue crabs were scarce, and that whooping cranes seeking food had wandered far from their usual territories.

In 2013, the judge ruled that Texas violated the Endangered Species Act and must regulate water withdrawals to adequately supply the refuge with freshwater. Almost a year and a half later, the appellate court reversed the decision, ruling that while cranes may have starved, water regulators were not liable. Biologists counted a record number of nests, 83, in Wood Buffalo National Park in 2014, yet the future of whooping cranes hangs in abeyance, turning on whether in southern Texas we are willing to share life-giving water.

Where in this continuum of devastating loss and painstaking restoration are red knots? Each summer six lineages of red knot from all over the world—more than 1 million birds—fly to the Arctic. Worldwide, knots aren't threatened with extinction, but their numbers are declining. There are signs of trouble on almost all the flyways—in Africa and Asia, in Europe and North America.

Knots flying to the Arctic from Australia's hot and humid Roebuck Bay and New Zealand's long and sandy Farewell Spit rest and replenish their reserves in the Yellow Sea, where, as more ports and petrochemical companies are being built in Bohai Bay, south of Beijing, the birds are squeezed onto fewer and fewer tidal flats. Shorebirds there are running out of room. If knots lose this essential stopover, their population will be in danger of collapse. In 10 years it has already dropped by half, to 100,000 birds.

A larger cousin of the red knot and a different species—the great knot—has also been hurt by the loss of its Yellow Sea home. One-quarter of its population used to fly to the Saemangeum estuary, the Yellow Sea's premier staging area for shorebirds. When construction of a 22-mile seawall there took away its stopover, the population dropped by 20 percent, catapulting the great knot onto the IUCN (International Union for the Conservation of Nature) Red List of Threatened Species.

The knot I'm following, *Calidris canutus rufa,* has already been listed as endangered in Canada and as threatened in the United States. Another red knot subspecies, *roselaari,* migrates along the Pacific, from the world's largest saltworks in Mexico's arid Baja, up through Willapa Bay and Grays Harbor in the Pacific Northwest, and on into the Russian and Alaskan tundra. Their numbers are so small, only 17,000 birds, that any loss along their flyway is a threat. Neither of the two populations that fly through Europe, *islandica* and *canutus,* with 450,000 and 400,000 birds each, is in imminent danger of disappearing, but their ranks are thinning. *Calidris canutus canutus* is a spectacular long-distance traveler, wintering in the tidal flats in Mauritania's Banc d'Arguin National Park at the edge of the Sahara or in the mangrove-rimmed river mouths of Guinea-Bissau's Bijagós Archipelago, tropical wintering grounds for 3.5 million birds. A group that once wintered in the saltwater tidal flats in Cape Town, South Africa's Langebaan Lagoon, about 12,500 birds, has all but disappeared.

High numbers can vanish in a flash. The Eskimo curlew, a small bird with a curved bill, once blitzed through Labrador on its way south, gorging on blueberries before turning out to sea and down to the pampas of Argentina. In the spring, Eskimo curlews flew to the Arctic through the Great Plains, filling up on grasshoppers. An observer in Labrador in 1860 describes a flock "perhaps a mile long and nearly as broad," whose passing sounded like "wind whistling through the rigging of a ship" or the "jingling of many sleigh bells." The Eskimo curlew was also called the dough-bird, "to denote an extremely

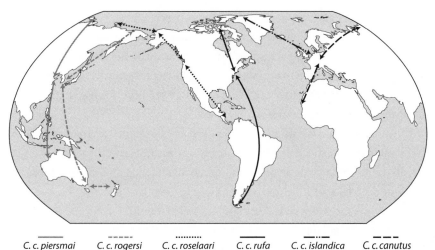

C. c. piersmai   C. c. rogersi   C. c. roselaari   C. c. rufa   C. c. islandica   C. c. canutus

Red knots worldwide (map by Bill Nelson, modified and adapted from an article by Buehler, Baker, Piersma, *Ardea* 94, 2006).

fat and delicious fowl." Hunters shot thousands in a single day, and farmers and settlers turned their grassland home into fields of wheat and corn. A population that "numbered at least in the hundreds of thousands" was decimated in a few decades. It hung on for years, a few birds here and a few birds there, the last confirmed sighting when I was a young girl.

We've killed too many Eskimo curlews, spoon-billed sandpipers, ivory-billed woodpeckers, and whooping cranes. We've cleared away their homes by cutting old-growth forest, plowing the prairie, and walling off the sea. I am haunted by these birds. As I travel along the knot flyway in South and North America, I meet many people determined to prevent the knot from declining further, to restore what we are in danger of losing. They face loss holding the possibility of repair. The stakes are high: as the knot goes, so go many other shorebirds.

Tierra del Fuego is the first rung on a ladder that takes red knots up to the Arctic. Underlying the tranquility and isolation of the empty, windswept beaches of Bahía Lomas lies a vague but persistent disquiet. Along the Strait of Magellan, a place stunningly beautiful and difficult, the seasonal arrival and departure of shorebirds has taken place within a tumultuous history of human arrivals and departures, defeats and conquests, rises to glory and subsequent declines. Spanish conquest and colonization of southern Chile came neither quickly nor easily. One explorer, Juan Fernández Ladrillero, staked a claim for Chile in 1558 at a place he called Posesión. The name stuck, but possession proved difficult. The Spanish could barely navigate the strait, let alone make it their home.

In 1584 Pedro Sarmiento de Gamboa brought 300 people to settle on the strait's northern bank. One of their ships foundered on the treacherous shoals, sinking its guns and provisions. Rough winds and storms rebuffed not one but two relief ships. A few people eked out a living foraging on shellfish, but most starved. Thomas Cavendish, coming upon the miserable scene three years later, promptly renamed the settlement Puerto Hambre, Port Famine. Another 300 years would pass before the Spanish established, at Punta Arenas, a colony that would last. During it all, knots were making their winter home on Bahía Lomas and Bahía Posesión.

So much has come and gone since the founders of Punta Arenas—a Buenos Aires banker, a Portuguese sailor, and a Lithuanian accountant—built an empire from a flock of imported sheep. José Menéndez, José Nogueira, and Mauricio Braun created Chile's largest ranching operation, raising millions of sheep and shunting aside millions of guanaco that once roamed the Patagonian steppes. Meat-packing plants, tanneries, and rendering plants sprang up along the strait. The Menéndez mercantile fleet grew to meet the rising demand for meat and wool, shipping thousands of tons of cargo through the strait's rough waters to Liverpool or up the Pacific coast to Valparaiso.

When the Panama Canal opened in 1914, this "golden age" abruptly ended. On the way to Bahía Posesión to meet Morrison, I passed what was left of the *Amadeo,* the ship that launched the Menéndez fleet, rotting on the San Gregorio beach, just before the First Narrows. All but a few of the meat-packing plants are closed. Even the sheep population is declining. Overgrazing the harsh land, sheep have reduced 30 percent of the Patagonian steppe to desert. In the last 50 years their numbers have dropped by two-thirds as unpalatable shrub replaces grass. The estancia at San Gregorio, part of the Menéndez wool empire, is a ghost town. There are other ghosts. As the new settlers carved out their estancias, they uprooted and extirpated the indigenous Selk-nam, who'd lived along the strait for thousands of years. Of their passage I saw little trace.

Knots flew into Bahía Lomas each October and left in March as the light faded. Was their livelihood affected by the decline of guanaco and the rise of sheep husbandry? Did one replacing the other alter the ecology of the beach, the availability of food? Did packing and rendering plants and tanneries alter water quality or nutrients in the strait in a way that mattered? If the numbers of knots were much larger when the Spanish arrived, we may never know. Their history is blurred by the passage of time, and Morrison and Ross made their baseline estimates of bird populations long after the guanaco had declined, long after the wool empires had begun to fade.

The wool industry declined, but another boom began, the rough land having more to yield. On December 29, 1945, after 40 long years of exploration, drillers finally hit oil on Tierra del Fuego. They named the gusher Manantiales, or Springhill. The Chilean national oil company began to extract the oil, and by the close of the decade, a new town, Cerro Sombrero, had risen from the dusty pampas. Built to accommodate ENAP's workers and their families, Cerro Sombrero is a planned community, inspired by Le Corbusier and modernist architect Oscar Niemeyer's Brasilia.

Heralded as part of "la realización de una utopía moderna en el fin del mundo," a modern utopia at the end of the world, the town, built for 200 families, represented the aspirations of a modern Chile. Enormous orange, yellow, and black panels made up the façade of the village's new theater—a giant Mondrian on the plains. A large sports complex on the town square sat atop a broad staircase. Its design, three glass parabolas, looks like three ships sailing through the landscape. Inside were a gym, botanical garden, and heated swimming pool. On an outdoor chess board, players walked from square to square, moving pieces made from real drill rigs. In 1971, Presidents Salvador Allende and Fidel Castro came to Cerro Sombrero to celebrate its success, and in 2008 Chile recognized the town as one of the country's most important architectural achievements. By the 1960s Cerro Sombrero was a bustling village in a region supplying three-quarters of Chile's petroleum needs.

The oil boom came and now it's gone, the reserves exhausted, the wells running dry, the work winding down. When we arrived, Cerro Sombrero looked tired. The windows of the sports complex were broken, the swimming pool closed, the botanical garden dusty, droopy, and dilapidated. The school, built to accommodate 200 students, had 80. The small supermarket sold frozen pizza, wine in boxes, ancient carrots and pears. The cavernous central dining room, where we usually ate, was mostly empty. Oil extraction gave life to Cerro Sombrero, but the present will soon be past, and no one is sure what will come.

The Strait of Magellan, with its high winds, racing currents, and sweeping tides, still challenges navigators. In August 1974 the supertanker *Metula,* owned by Royal Dutch Shell and en route from Saudi Arabia to Chile, grounded just west of the First Narrows, dumping more than 50,000 tons of crude oil into the strait. The spill, then the second largest in the world, surpassed that of the *Exxon Valdez.* Wind and currents coated between 125 and 150 miles of beach with a thick emulsion of oil that ranged between 50 and 200 feet wide. According to a report prepared by the science advisor to a strike force sent by the U.S. Coast Guard, "Almost everything that would have been necessary for control of the oil spill was lacking or totally non-available." A limited amount of dispersant was flown in two weeks after the spill. There was no equipment to apply it. Nor was there a major cleanup. A few front-end loaders couldn't remove what would have required 12,000 loads in dump trucks.

A brief survey of 25 miles of beach found dead and dying cormorants, penguins, terns, ducks, albatross, plover, whimbrel. The science advisor, Roy Hann from Texas A&M, wasn't able to visit beaches in Bahía Lomas and Bahía Posesión where oil had also washed ashore. Whether the spill oiled red knots and Hudsonian godwits will never be known. The main study sites were west of beaches that Morrison and Ross would only later identify as a central home to red knots.

More than 30 years later, *Metula*'s signature was still in the strait. Oil-covered marshes were still mostly bare of vegetation, but seeds had begun sprouting in animal hoof prints. An 1,800-foot pavement of asphalt on the beach had slowly begun eroding. Tankers are double hulled now, but the strait is still treacherous. In 2004, when a ship and a tug collided near the First Narrows, the area still lacked the means to mitigate an oil spill. Three days after the spill, a small team of contractors began manually removing oily kelp and shingle from the beach. Eighty-eight percent of a colony of cormorants died.

If oil spills in the strait reach shorebirds on the remote beaches of Bahía Lomas, the birds may carry the consequences along the entire flyway. Oiled feathers, no longer waterproof and no longer providing insulation, leave them vulnerable in chilly winds. Knots need to be well nourished for their long flights north from Tierra del Fuego: taking time to preen their oiled feathers, they may eat less. Ingesting oil as they forage, they can become anemic, losing oxygen-carrying red blood cells that provide needed stamina for long-distance flights. Failing to gain sufficient weight, they may delay their departures, setting off a domino effect that will accompany them all the way to the Arctic. Black-tailed godwits in poorer, inland winter quarters in England leave later for their breeding grounds in Iceland, and breed less successfully than those wintering in more fertile coastal marshes. Ten years after the *Exxon Valdez* oil spill, female harlequin ducks wintering in contaminated sites don't live as long as those wintering along shores untouched by oil. Anemia can impair a bird's egg laying. The damage isn't limited to birds consuming contaminated prey: after an oil spill off the coast of Spain, hydrocarbons appeared in the eggs of peregrine falcons that had consumed shorebirds.

It's hard to ascertain how well Chile is equipped to deal with a spill in the strait today, or with the kinds of oiled birds that Morrison and Harrington found in Argentina during their road trip. While I was in Chile, the national park of Torres del Paine, famous for its glacial lakes and soaring peaks, was besieged by raging wildfires. The woods were charred, the grasses scorched. I saw a burned ranger station, everything gone but a porcelain bathtub that stood untouched amid the ruin. Animals huddled in patches of green. The park burned for two weeks: the air was heavy with smoke. If Chile had difficulty controlling fire in Torres del Paine, the crown jewel of its national parks, attracting thousands of visitors every year, how would it handle oil spills in an equally remote but far less heavily traveled area?

On the beach at Bahía Lomas, I met Oscar Oyarzun, mechanical engineer and regional administrator for the Chilean national oil company. A colleague and he had made, without benefit of an ATV, the long trek through the mud to watch knots flying in. "Una sensación maravillosa," he wrote me later, "el sonido del inicio del vuelo en la inmensidad de Bahía Lomas es inolvidable" (In the immensity of Bahía Lomas, the sound of the start of the flight is a wonderful sensation—unforgettable). ENAP is committing to ensuring the safety of commissioned and decommissioned pipelines and platforms, designing plans to rescue and treat oiled birds, and updating protocols for dealing with oil and chemical spills. While I was in Cerro Sombrero, a lone red box perched on a river bank held equipment to be used in case of an oil spill. It wasn't much. Perhaps, before the next spill, the resources to carry out ENAP's commitment will have been allocated. Espoz and her colleagues are inviting the larger public to consider how much the isolated beaches of Bahía Lomas mean to shorebirds. Inspired by a great journey taken by a small bird, they hope to inspire others to become its guardians as well. Working to realize a long-held dream that birds and humans can safely share this shore, they seek additional protection for both the birds and the beaches—protections built into whatever the future may bring here, whether it be natural gas, tourism, ranching of one kind or another, or something not yet imagined.

Nearby in Río Grande, Argentina, where Harrington and Morrison found their largest concentration of knots more than 35 years ago, the birds are disappearing. By 2012 only 300 remained—a staggering loss of 94 percent. Río Grande, growing out toward the sea and the edges of the Río Grande River, crowded out the birds, leaving them fewer places to roost. They feed amid congestion, constantly interrupted by the commotion of off-road vehicles, dogs, and people. Forced to take flight repeatedly, they lose precious refueling time. Minutes lost during one ebb tide on one day accumulate into hour upon hour as the season continues. So many times I'd walk the beaches at home, unconsciously flushing flocks of sandpipers at the tide line, taking pleasure as they circled out over the water and then landed farther down the beach, never thinking that disturbing them might make a difference.

Not only did knots at Río Grande have less time to eat, their food became less plentiful. The area's exceptionally abundant tiny clams and mussels drew knots to this distant place at the end of the world, but in 2008 this plenty diminished. Clams and mussels were significantly smaller, mussels were fewer, and the clams were infected with parasites. The thousands of birds Harrington and Morrison saw seemed plentiful, but today's population of only a few hundred raises the possibility of their disappearing altogether.

Along the strait the number of knots is dropping as well. By 2013 Morrison and Ross's original mother lode of 41,700 birds was down more than three-quarters to a mere 9,900 birds. Perhaps the losses here are a legacy of oil extraction in the strait and oil spills from ships, an unseen, unmeasured burden that accumulated year after year. Perhaps a stress thousands of miles away on another beach in another bay is being manifested here. Or perhaps a change in the quality of the water diminishes its fertility. Bahía Lomas, like the Aransas National Wildlife Refuge, may be susceptible to human actions taking place well outside its borders. Espoz and her colleagues, continuing their lonely work, may uncover the reason. Whatever it is, in Bahía Lomas , in what may be the last refuge for red knots on Tierra del Fuego, it may prove difficult to climb a ladder when the first rung is cracking.

Three

THE URBAN BIRD AND THE RESORT

Río Gallegos and Las Grutas

In the long summer days and short nights along Bahía Lomas, red knots can feed on two incoming tides, but as summer ends, the light fades, the rich stores of clams and mussels dwindle, and the birds head north. By March, the end of summer in the Southern Hemisphere, the sky over Bahía Lomas fills with departing knots. A very few take a short hop across the strait to Río Gallegos, Argentina. Red knots may no longer find a home in this growing city, but I wanted to make a short stop here to see where they once lived.

The wind is whipping when we arrive. It seems even more intense than in Bahía Lomas. At the seawall, oystercatchers, white-rumped sandpipers, and a whimbrel hunker down in a tiny bit of marsh littered with bottles and cigarette butts. They attempt, infrequently, to stagger across the sand. No knots. I'm hoping for another glimpse of the elusive Magellanic plover, a bird that is easily camouflaged in the gray mud and gravel of a beach but for its bright pink legs. If we are to see one again, it will be here. More Magellanic plovers live in Río Gallegos than anywhere else: two flocks of over 100 birds each winter along the estuary. I know I won't see chicks or, unusually for shorebirds, parents regurgitating food into the beaks of their young, but possibly I might catch one's shuffling gait or see it fly by. But not many birds are aloft. We're doubled over in the wind and can't see much of anything.

We go to the dump. Trucks drive in and out through dark, rancid muck. In the dump in Ushuaia, at the southern tip of Argentina, taxis wait for birders who have come to check out the white-throated caracara. In Mark Obmascik's book *The Big Year* (chronicling an intense competition to break the record for the highest number of birds seen in one year), a contestant seeks the Tamaulipas crow at what cognoscenti call the Tamaulipas Crow Wildlife Sanctuary—known to everyone else as the Brownsville, Texas, dump—so popular with birders that a sign points the way to the best sighting. Gulls are the primary avian visitors at Río Gallegos's dump, which is burying the salt marsh.

Not all that long ago New York City piled its trash onto the salt marsh on Staten Island. The Fresh Kills Landfill, formerly the world's largest landfill—more than two and a half times the size of Central Park and covering 11 percent of Staten Island—was open until 2001. Now that the city is building a world-class park there, the salt marsh is growing back. With the fresh garbage gone, thousands of seagulls are seeking their meals elsewhere, and other birds are coming back: Lapland longspurs and snow buntings; egrets and herons; teal and pintail duck; greater yellowlegs and spotted sandpiper; cooper's, red-tailed, and sharp-shinned hawks; and bald eagles. Knots once visited the shores of Staten Island, but over the years their numbers declined. Now, perhaps, there is room for them again. James Corner, architect of New York City's High Line, is designing the park at Fresh Kills. As the sea rises, it can't be completed soon enough: the restored marsh fronted the surge of Hurricane Sandy, sparing nearby neighborhoods.

In Río Gallegos, not that far from the dump, roads and newer houses encroach on what is left of the marsh. The city is bursting at its seams. In 1960 the city claimed 14,400 residents. By 2010 the population had jumped by more than 750 percent, to 110,000. This coast is home to birds as well as people: 20,000 migrating and local shorebirds take shelter here, including red knots. When the birds arrive in Argentina, they have a new name. They are no longer *playero ártico;* the Argentines call them *playero rojizo* for their red breeding plumage.

Red knots in Río Gallegos have declined sharply. In 2006, 3,000 were seen feeding in the estuary's broad tidal flats. In 2007 the count dropped by half. In 2009 it was 600, and in 2010 only 110. Red knots may be losing their home here.

Into this deteriorating coastline stepped two scientists, Silvia Ferrari and Carlos Albrieu. They met 30 years ago at the University of Cordoba in an invertebrate zoology class. They are in love: with each other, with shore-birds, and with the Río Gallegos estuary. Studying the marsh, they saw it slipping away, sold off for house lots, heaped with garbage. They analyzed and documented the extensive loss—360 acres, or 40 percent, gone by 2003. They didn't stop there. In the United States, the relationship between science and advocacy can be complex and contentious. In a 2011 workshop held by the American Association for the Advancement of Science and supported by the National Science Foundation, participants discussed whether scientists advocating for specific laws or policies on the basis of their expertise and work compromised their objectivity, legitimacy, and credibility, tainting their work and recategorizing them as "lobbyists." They discussed whether, as concerned citizens of a world increasingly complex and difficult, they have the right—and perhaps the responsibility—to engage in public policy and advocate for the public interest. American scientists can face wrenching challenges by well-funded, powerful indus-tries doubting and undermining their findings.

For Ferrari and Albrieu, the choice was clear. The rich marsh provides a significant and economically valuable buffer from floods and storms. Essential to the livelihood of thousands of birds, it also presents an economic opportunity for controlled tourism and educational recreation. They believed a healthy salt marsh could be a source of great pride for Río Gallegos. Further, the city needed the marsh. As more and more land was filled, flooding increased, forcing the city to build an embankment to keep the water out. For Ferrari and Albrieu, the public interest trumps the financial interests of a minority.

Working with the municipal government and other allies, Ferrari and Albrieu made a prodigious number of changes in an astonishingly short period of time. They assisted Río Gallegos in creating two protected areas, one along still-undisturbed coast outside the city and another downtown, the Urban Coastal Reserve, which prohibited further building in the marsh. They convinced the city to create a Department of Environmental Management. They and the conservation and education group they helped found, Ambiente Sur, worked with the city, the Manomet Center for Conservation Sciences, and the U.S. Fish and Wildlife Service to build a visitor and education center at the seawall, set up bird-viewing areas for the public, and wrote a birding guide.

Shrinking tidal wetlands in the Río Gallegos estuary may have forced knots from their home. In the case of another bird that nearly slipped away here, the problem originated outside the city. The sudden and unexpected decline in the hooded grebe, from 2,500 birds to 400, originated, surprisingly, not in the urban estuary but along inland lakes in the Patagonian steppes where the birds breed. Together, an accumulation of injuries led to their near demise. Humans introduced rainbow trout into the remote lakes and mink onto the arid steppe. Kelp gulls, amply fed on garbage, grew in number. They all raided the nests of hooded grebes. Sheep grazed the lake edges, cropping the vegetation, leaving nests exposed to exceptionally high winds. With less snowfall, water levels are dropping in the lakes. Ambiente Sur is as dedicated as Ferrari and Albrieu. The group documented the hooded grebe's falling numbers, determined the cause, and found remedies. Working with other conservation organizations, the organization put the hooded grebe on the IUCN Red List of endangered species, and in 2013 created a national park, Park Patagonia, to protect the grebes. Now, full-time rangers protect the nests and stave off predators.

The work continues as Albrieu and Ferrari look to their next—and monumental—task: raising funds to move the dump. I've no doubt that under their aegis this will happen. Listening to Ferrari and Albrieu, I wonder when, or whether, they sleep.

As many as half the knots leaving Tierra del Fuego fly 900 miles up the Argentine coast to the beaches of San Antonio Oeste and nearby Las Grutas, a village built at the edge of limestone cliffs rising from the sea. Waves cut large caves into the cliff base, giving the village its name. The water, deep blue and warm, some of the warmest along the Argentine coast, reaches a pleasant 75 degrees at the height of summer. We arrive at this critical stopover on the knots' northbound flight in early March. Knots are arriving as well. I've put away my heavy jacket and sweaters, traded my sneakers for sandals. Knots, unable to switch to lighter feathers and lacking sweat glands, cool off their own way: heat from their bodies escapes as blood flowing through their skinny legs increases. Their "hot legs" reduce their body heat by as much as 16 percent.

Out on the mudflats, a feast awaits them. Millions of years ago dust and silt blew off the Patagonian steppes and swept down through the valley of the Río Negro, settling in the bay. Over time it sank and compacted. Held together by minerals from the shells of animals that lived and died in the bay, it built up into a broad, slippery shelf—the restinga. Exposed by the ebbing tides, restinga, softer than rock, harder than mud, houses dense beds of *mejillines,* tiny mussels, easy for knots to pry loose and small enough to swallow whole.

Sanderling nab tiny mollusks and crustaceans as they chase waves sliding down the beach. Oystercatchers picking their way through shellfish beds thrust their bills into partially opened oysters and mussels, sever the muscle, and extract the meat. Knots are mollusk lovers as well, but unlike oystercatchers, they gulp down smaller mussels whole, crushing the shells in their muscular gizzards before digesting the meat. Out on the restinga, the hunt begins, but the atmosphere, unlike the quiet, empty beaches of Bahía Lomas, is hectic.

Movimiento, action, characterizes Las Grutas, on the beach and in town. We wake to the sound of backhoes and excavators. Outside our apartment, construction workers are clearing scrub, hewing new roads from scrabble. The dry steppe is being carved into house lots. A water truck rolls by, spraying down the dust. We prepare the first of several membrillo and bread breakfasts. The quince that Magellan brought across the sea to feed his men today comes chunky or plain, sweet or sour, as liquid, paste, or other gradations of firm. We cut slices from a large and quivering block, slather it over homemade bread, and gulp it down, then walk into the burgeoning town.

When Argentina defaulted on its debt and the economy collapsed in late 2001, and the country had four presidents in rapid succession, people who'd vacationed abroad stayed closer to home. The summer population in Las Grutas soared. San Antonio Oeste, of which Las Grutas is a part, isn't fully managing its explosive growth. Apartments and hotels hug the road along the cliffs, a little too close to the edge. A piece of road has collapsed, along with a *bajada,* path, leading down to the beach. We pick our way through the rubble. When the tide is out, it's a long walk to the water. Some complain that it's too long. To accommodate bathers, the city is blasting the restinga into large *piletas,* shallow pools, to hold seawater as the tide ebbs. I can't see to the bottom, only a few feet down. The edges of the pool are slimy, the water murky. Heavy machinery and noisy ATVs zipping along the beach crush the mejillines and scare the birds. Only a few years ago, if you wanted to ride an ATV onto the beach, you rented one of 14 in the entire town. Now more than 1,000 crowd the streets and beaches of Las Grutas, almost all privately owned.

In this hubbub, knots need to find an abundance of food and find it quickly. Mussels are mostly shell, ballast to be jettisoned, but they are plentiful. They are tiny. In Las Grutas, knots prefer mussels no more than half an inch long, swallowing them one or two every second. Their need to feed is urgent. When night falls and they can no longer see mussels on the restinga, they seek their meals on other beaches and by other means. On the beaches near San Antonio's port, they probe water-soaked tidal flats with their beaks, using special sensors at the tips of their bills. These Herbst corpuscles detect pressure waves created around their prey, in this case, small clams.

Shorebirds refuel quickly, more quickly than most birds, consuming 80 percent of their average weight in food every day. Some knots in San Antonio Oeste double their weight before leaving on the next leg of their journey. The prospect of doubling my own weight in a month is both daunting and nauseating. In the movie *Supersize Me,* Morgan Spurlock stuffed himself at McDonald's, gained 24.5 pounds, and increased his body mass by 13 percent in a month. Knots can increase theirs by almost 10 percent in a day.

On a shoreline growing increasingly crowded, two women began speaking for birds and other wildlife whose homes were disappearing. In 1988, Mirta Carbajal, former member of the Argentine national volleyball team, moved to San Antonio Oeste. She'd traveled around the world four times with her husband, a master mariner, and when their first child was a year old, they moved to San Antonio Oeste, where he piloted large ships in and out of the port. Around the same time, Patricia González, who'd been living in Buenos Aires with her husband and children, also moved to San Antonio Oeste, when her husband, an architect, relocated. Both women are biologists. González's area of expertise is shorebirds. Carbajal likes animals that others fear—bats and spiders. About 30,000 people live in San Antonio Oeste, in the port, and in Las Grutas. The women quickly found each other and joined forces in an alliance that is changing the face of the region and the lives of those who live there.

When they arrived, construction of a chemical factory was under way. The factory, using sea brine, ammonia, and limestone, would be making soda ash—sodium carbonate. Soda ash is used to manufacture glass, as a common additive in detergent and toothpaste, and an ingredient in ramen noodles. It makes sherbet fizz. The plant in San Antonio Oeste produces enough soda ash every year to fill 2,000 railcars, and an equal amount of salt, a waste product intended to be thrown away in the bay. The salt wouldn't be disappearing, however. One of González's colleagues, a fisheries biologist, had found that the waters of San Antonio Bay are enclosed: fish larvae drifting out with the tide to the mouth of the bay returned. At first, these findings were unwelcome. González recalls that the biologist and his family, who lived in government-owned housing, were threatened with eviction from their home. To her, piling up so much salt and making the water saltier was unconscionable.

González and her colleagues persuaded the provincial government to designate San Antonio Bay a protected area. They demanded that the government disclose its study of how the plant would alter the bay. In the meantime, the people of Las Grutas successfully blocked an oil pipeline from being routed through the town, and González, Carbajal, and others revitalized a foundation, Inalafquen, to carry on their work. Its name means By the Sea. The women continued to insist that the chemical factory comply with existing laws. They insisted the sea was not a dump. Because the plant promised jobs, those who opposed it were harassed. Carbajal's car was scratched, graffiti scrawled on her house. Inalafquen formed an alliance with more than 50 organizations and institutions and demanded that the government examine the consequences of throwing waste into the ocean. According to Carbajal, the company then produced seven volumes minimizing the impact. A team of scientists analyzed them all. "It was *tocar el violin*, fiddling," Carbajal says. "Their words made no sense." The years rolled by. Eventually, the provincial government was persuaded that, in González's words, "the sea is nonnegotiable." The salty wastewater now evaporates in a lagoon lined with an impermeable membrane.

The soda ash factory sits at the end of a gleaming white salt road near a beach. Here, from time to time, knots abandon Las Grutas's skimpy mejillines for luscious sea worms. In good years, when the flats are full of worms, knots fatten up on this better meal. Who knows how dumping waste salt from the soda factory into the bay would have further altered the seafloor, altered salinity, and affected the lives of sea worms? A 2013 review of brine discharges from desalination plants, prepared for the California Water Resources Control Board, expresses concern about how communities of sea life are restructured and diminished when brine is added to poorly flushed waters. In the seabed near a desalination plant in Spain, the number and variety of sea worms declined within two years of the plant's opening. Seahorses, whose international trade is restricted under CITES, the Convention on International Trade in Endangered Species, live along the shallow, sandy bottom of San Antonio Bay. When the soda ash factory opened, the community of tiny animals living in the waters around the plant disappeared, including the seahorses. Today, a few are seen, but only occasionally.

If the plant had been allowed to dump hundreds of thousands of tons of salt into the bay, many more animals might have been lost. When Carbajal and González insisted that the sea was nonnegotiable, they were safe-guarding a world they couldn't fully describe, whose inhabitants they didn't fully know—a world that is essential to knots and other shorebirds.

The day González and I go to the port is scorching. We hope to find shorebirds escaping the hot dry wind on these more sheltered beaches. I've brought two large bottles of water, but when we stop for gas González purchases more water, supplementing it with Gatorade. I buy as much as I can carry. The port of San Antonio is busy—80 percent of fruit grown in Argentina's lush river valleys and bound for Europe, the United States, and Russia comes through here. By the end of the day, long after I'd gulped down the last of my water, I'd imagine quenching my thirst on the pears and apples, peaches and plums, nectarines and grapes on passing container ships. We walk and walk and walk in the hot sand and glaring sun looking for knots.

González didn't come to her expertise in shorebirds by choice. When she moved to San Antonio, the only local person her university in Buenos Aires approved as her thesis advisor studied birds. Therefore she studied birds, aided by her grandfather's old heavy binoculars and a black-and-white field guide. It was a rough start. In the beginning, even this expert once imagined that she'd seen birds, like the crested duck, when she hadn't. Her advisor suggested she draw birds she sighted or thought she sighted, but she couldn't draw and didn't own a camera. Digital cameras, cell phones, and bird apps didn't exist then. A fisherman and former mayor of San Antonio took González to find shorebirds, and over time she came to know them. On Oasis Beach north of Las Grutas, she found huge flocks of white-rumped sandpipers, and at a port beach a Magellanic plover spinning in the sand as it pecked for food. Near the soda factory, she found knots beginning to assume their rust breeding plumage. First she saw 20 and later 7,000.

She learned to band birds, trapping them at night with a mist net on the beach, only to be questioned by the police. "Everything was difficult. I'd be talking about *chorlito,* plover, but I'd be called *cabeza de chorlito,* scatter-brain. I wanted someone to come here and see what I was seeing." At an ornithological conference she attended in Ecuador, she met Guy Morrison, but since neither spoke the other's language, they couldn't converse. She met Theunis Piersma, a Dutch researcher whose French she did under-stand, and under his tutelage developed research protocols. She taught herself English and began to publish.

Biology has a rich inheritance from Tierra del Fuego and Patagonia. Captain FitzRoy's predecessor on the *Beagle,* unhinged by the dark, unnavi-gable channels in the Strait of Magellan, the violent gales, and the gloomy impenetrable forests, committed suicide. The new captain, putting together his crew and perhaps fearing the worst, sought a gentleman companion in his naturalist. Young Darwin, who'd rejected his father's recommendations to make a career in medicine or the church, eagerly signed on. While FitzRoy undertook hydrographic work, Darwin explored the shore. From his obser-vations and collections would emerge his ideas about evolution, adaptation, and natural selection.

Darwin's travels through the pampas along the Río Negro, not that far from San Antonio Oeste, and further north along the cliffs of Bahía Blanca proved particularly fruitful. "Having heard of some giant's bones at a neigh-boring farmhouse on the Sarandis, a small stream entering the Río Negro," he writes, "I rode there . . . and purchased for the value of eighteen pence the head of the Toxodon." The bones were cheap, their intellectual value immeasurable, ultimately leading Darwin to realize that the succession of animals over geological time pointed to a resolution of that "mystery of mysteries," the origin of species.

Darwin was among the first to collect the massive bones of toxodon and other giant animals once living here. The elephant-sized megatherium; the ground sloth mylodon, which had reared to a height of 10 feet; the bony-plated glyptodont, about the size and heft of a Volkswagen Beetle; and the long-necked macrauchenia: all are emissaries from a world long gone. Darwin walked amid their skulls and skeletons, scattered in an area about the size of a volleyball court along an ancient gravel beach in Patagonia. On another beach he found shells, a historical perspective on our contemporary diminished ocean: Darwin's oysters were a foot wide.

He saw, in smaller, living animals, glimpses of these dead giants: in wild guanaco roaming the pampas, he saw macrauchenia, an extinct llama; in capybara, large grass-eating, piglike rodents, he saw toxodon; in tiny armadillo, which Darwin ate for dinner, he saw armored glyptodont; in small tree-climbing sloths, he saw mylodon and megatherium. Writing in his journal, Darwin was already considering the possible larger meaning of his findings. "This wonderful relationship in the same continent between the dead and the living, will, I do not doubt, hereafter throw more light on the appearance of organic beings on our earth, and their disappearance from it, than any other class of facts."

In Patagonian birds, Darwin saw evidence of another idea central to his understanding of evolution: that species are linked not only through time but also across geographic space. The gauchos told him about two different rheas—larger birds north of the Río Negro and smaller ones, much more rare, to the south. One night he accidentally ate the evidence of this important idea. A rhea had been shot for dinner, he recalls, "and I looked at it, forgetting at the moment, in the most unaccountable manner, the whole subject. . . . It was cooked and eaten before my memory returned." But not fully eaten. "Fortunately the head, neck, legs, wings, many of the larger feathers, and a large part of the skin had been preserved; from these a very nearly perfect specimen has been put together and is now exhibited in the museum of the Zoological Society." The new species, despite its awkward entrance into the world of zoology, was for a time given Darwin's name.

Patagonia's rheas continued to fuel his ideas. He hunted them with the gauchos, noticing that when they are chased they break into a run, "expand their wings, and like a vessel make all sail." He watched other flight-less birds in Patagonia and Tierra del Fuego: penguins and steamer ducks that, despite rapid flapping, didn't lift off. Their wings were too small and weak. Once called racehorses for their speed—they've been clocked at 10 knots—steamer ducks leave a bubbly trail as they churn through the water. In his observations that wings worked as paddles for steamer ducks, as fins for penguins, and as sails for flightless rhea were seeds of his later ideas about adaptation and evolution. By the time he wrote *On the Origin of Species,* these three very different wings would "serve to show, at least, what diversified means of transition are possible."

From Darwin came the theory of evolution and natural selection. Gregor Mendel, cultivating green and yellow peas in an Austrian monastery, founded the science of genetics. The idea that genetic mutations occur randomly, that those who survive are those with mutations best adapted to a particular environment, became a cornerstone of biology. DNA was destiny. Jean-Baptiste Lamarck, predecessor to Darwin and professor of zoology at the Musée National d'Historie Naturelle, proposed that organisms actively respond to their environments, acquiring adaptations they pass on to future generations. Giraffes, he thought, strained their necks to reach higher leaves and thus gave birth to offspring with longer necks. The idea seemed preposterous. Lamarck died in poverty, buried in a rented grave, his views denounced.

Giraffes did not acquire long necks in the way Lamarck imagined, but his underlying idea is undergoing reconsideration as scientists in the burgeoning field of epigenetics find that animals developing traits in response to their environments can pass on those traits without changes in their DNA sequence. Male rats, becoming obese eating fat-laden food, produce daughters with tendencies to diabetes; pregnant rats exposed to bug repellants, jet fuel, and pesticides give rise to daughters, granddaughters, and great-granddaughters with ovarian diseases; rats exposed to endocrine disrupters also transmit disease through generations; and chickens whose days alternate irregularly and unpredictably with night develop new habits of foraging that then appear in offspring whose days and nights follow a normal rhythm.

Animals once thought to passively await random genetic mutations in the face of environmental challenges are now also understood to alter their behavior in ways their offspring can inherit. For more than 20 years, González's teacher Theunis Piersma and his colleague Jan A. van Gils have been documenting the exquisite flexibility, described by scientists as phenotypic plasticity, of red knots facing the rigors of long-distance migration. In 2014 Piersma was awarded Holland's prestigious Spinoza Prize for his research on shorebirds. His studies of knots led him to suggest that differences among the six lineages of knots—differences in feeding habits, length and timing of migration, coloration—evolved too quickly to be the result of random gene mutations and natural selection. Instead, new traits emerge as the birds, deeply attuned to their surroundings, respond, and their offspring inherit the responses.

Piersma and van Gils are cataloguing the astonishing and reversible body changes knots undergo to survive days on the wing with neither food nor rest. Not only do they double their weight to stay aloft for 2,000, 3,000, or 4,000 miles at a time, they also burn the energy, lose the weight, and then gain it again at the next stopover. On the ground they expand their gizzards by as much as 50 percent to accommodate their diet. Knots eating soft, easy-to-digest, high-energy sea worms or shrimp can get their calories with smaller gizzards. Harder-to-digest, lower-quality mollusks require larger ones. When fishermen took cockles from Europe's Wadden Sea, and knots no longer had the food they needed, their gizzards grew even larger, allowing them to consume whatever they could find. When the birds are ready to fly and no longer need to eat, their gizzards begin to shrink again.

Knots fear peregrine falcons, and Piersma and van Gils have determined that when they are roosting in the Wadden Sea off Holland, their muscles adjust. If they roost on a sandbar with an open view of approaching falcons, their chest muscles, relative to their body mass, are smaller. If they roost on an island closer to their food but where falcons launch surprise attacks from a nearby dyke, they need to make a quick escape. Their pectoral muscles, relatively larger, are needed to help the birds execute sudden, tight turns in an erratic flight display intended to outmaneuver the raptors.

Today's rapidly disappearing tidal flats may test the flexibility of even these highly adaptable birds. Darwin, looking at Patagonian skeletons of giant animals now extinct, wondered what could have exterminated them, asking with uncanny prescience what role humans and a changing climate may have played in their demise. Today, more than 150 years later, González,

Carbajal, and other biologists along the entire length of the migration route are asking the same questions. Concerned that knots, flexible as they are, may not be able to withstand the collapse of their homes, they are determined to provide safe haven for shorebirds along the flyway.

Far out along San Antonio's port beaches, past the aromatic plants growing in the dunes, past a colony of grunting sea lions, around a point away from the channel where big ships are coming into port with the tide, González and I find oystercatchers, black-bellied plovers, ruddy turnstones, a few whimbrels, and 350–400 red knots. We stand, waiting. Hudsonian godwits fly in. The knots roost near the water, their heads peeking over a low ridge of sand. They are looking at us. We spend the next few hours quietly approaching, ever so slowly, only a few steps at a time. When the birds get skittish, we wait. If we move too quickly, they take fright and fly off, and we must begin the long walk again. I feel faint in the hot sun.

Finally, we're close. Researchers along knot migration routes have been banding birds, attaching colored flags to their legs, a different color for each country: orange for birds banded in Argentina, blue in Brazil, red in Chile, green in the United States. The flag's unique number/letter combination identifies individual birds. Fonts for the flags were chosen, after trial and error, for easy viewing, but reading them accurately takes skill. "People think it's easy," González comments, peering into her scope, "but the brain tricks you, restructuring what you see. You must read letters and recognize their shapes. And the colors change with the light." I'd see that later in Florida, when a knot took two steps and its green flag seemed to turn orange. After years of practice, though, González is confident. We find a flagged bird—L6U. Tracking flagged birds enables scientists to identify stopovers, estimate their population, and discover how many survive from year to year. Bandedbirds.org has collected information on 15,500 knots.

It's March in San Antonio Oeste. In May someone will see L6U on a beach in Delaware Bay. Another person has seen L6U on Hog Island, Virginia. In the next few hours, González will see birds flagged in Río Grande or San Antonio Oeste that over the years have been seen thousands of miles away. There's comfort in seeing the same birds return year after year—the numbered flag proof that tiny individuals can undertake immense journeys and return home.

The sun is setting. Dolphins swim by in the dusk. González is oblivious to the passage of time. Back near the sea lions, two young rangers, Gimena Mora and Amira Mondado, wait for her, drinking maté. Early in their jobs, they literally followed in González's footsteps, placing their feet in her tracks as she taught them to approach the birds slowly, patiently, quietly, and inconspicuously. It was love at first sight. Funding for their jobs is always uncertain, but each year they return, inspired by Carbajal and González's passion and dedication. They are studying marine biology and ecological tourism, and hope to work in these fields. Where they patrol, the beach is less disturbed, and sea lions bring their pups to rest on the sand. A four-wheeler approaches. Mondado greets the passengers, enthusiastically answers all their questions, and explains how they can help birds and sea lions by staying back. A couple with a child walks by. Mora shows the young boy how to use her binoculars.

Their positive approach doesn't always work. A couple drives across the sand and up to the sea lions, ignoring the rangers' calls and gestures. They blow their whistles. The man, angry, says he'll come back in a boat, and they can't stop him. He leaves, uttering a last remark in slang no one will translate for me.

"Strong science is the foundation of all our work," Carbajal had told me earlier, "but people come here for the sun. In the 1980s, people weren't watching whales, but now Europeans are coming to watch whales. We can create the same interest in birds. People living here can be proud of Las Grutas and San Antonio Oeste as a critical home for migrating birds." Instilling that pride is long, slow work. A large banner hanging over the road into Las Grutas depicts a beautiful woman in a bathing suit and the words "Las Grutas, ambiente de emociones"—environment of emotions. "We have to reach people through their emotions," González says, "but we are scientists. How do we do this?"

Silvana Sawicky, a young woman who now works at Inalafquen, grew up inland in the hills of Bariloche, and although she had often visited her grandmother in San Antonio Oeste, she had never seen a shorebird until she met González. Analyzing people's feelings about birds, she realized that while they supported conservation and shorebird protection in the abstract, they made little connection between their own behavior riding ATVs on the beach and walking their dogs, and the vulnerability of shorebirds. As

González explains, "People know shorebirds have problems, but think birds can just fly away. If they disturb a flock, the birds are free to go elsewhere. They don't see that birds need our help to survive."

And so Carbajal and González and those working or volunteering at Inalafquen appealed to people's hearts. With substantial support from Manomet, they raised money for a visitor center, Latitud Vuelo 40, where people can imagine they are flying with knots. From a restaurant with big picture windows overlooking the beach, they can eat *pulpito*, octopus, from the restinga, and watch real birds. Anahí Valverde and Horacio Garcia, who hold a long-term lease on the property, named the restaurant Jahuel—Mapuche for well, or eye of water. They've built an adjacent museum with many beautiful artifacts documenting the history of the Tehuelche and Mapuche people who, along with the birds, have long sought shelter and sustenance here.

Under the auspices of the international conservation organization Rare, Inalafquen undertook intensive outreach, inviting people to learn and care about knots. During the campaign, public awareness of knots and their need for protection rose by more than 100 percent. Today, the red knot is a symbol of San Antonio Oeste, regional buses are painted with red knots, and Inalafquen's mascot, Fabien—a human-sized red knot dressed in a scarf, flying cap, and goggles and named after the pilot in Antoine de Saint-Exupéry's novella *Night Flight*—regularly visits children in schools, participates in flash mobs, and leads dances on the beach.

Inside the exhibit is a quote from Saint-Exupéry's most famous book, *The Little Prince*: "On ne voit bien qu'avec le cœur. L'essentiel est invisible pour les yeux"—the most beautiful things in the world cannot be seen or touched. They are felt with the heart. Also in the exhibit is a quote from wildlife biologist Robert Gill. Gill, who studies the remarkable long-distance navigation abilities of shorebirds, has shown, along with his colleagues, that migrating bar-tailed godwits take advantage of winds and weather systems that span entire ocean basins—a phenomenon previously unrecognized and so far inexplicable. Gill believes that at any given time, these birds actually know where they are relative to where they are headed, and he notes, "Humans are also a flock; do we know where we are going?"

González and Carbajal have both received Manomet's Pablo Canevari Award, given every two years to a person or organization in Latin America for outstanding commitment to shorebird conservation. In San Antonio Oeste, winners of the high school beauty contest aspire to be biologists or rangers protecting shorebirds. Young women return to the visitor center to work year after year, when they can be paid, and often when they can't, to build exhibits, read flags on returning birds, and teach visitors about the importance of mejillines and beaches for migrating birds. Gabriela Mansilla remembers when González first took her to the beach and taught her about the knots. "Now I take people to see the birds and watch them discover a world they have never seen before."

Each year, students from 375 miles away raise money to spend a week with González on the beach, banding and resighting red knots. Their response is visceral: "Feeling such a small life in your hands," one student, Cande Lorente, tells me, "you feel the importance of protecting it." Emi Suarez describes the group as "the godparents of H3H," a knot they banded. They eagerly follow H3H who, fattened on mejillines, worms, and clams, left San Antonio one spring on a nine-day, nonstop, 5,000-mile flight to Florida. "To hold a red knot and feel its beating heart," Maria Belén Pérez, another student, tells me, "is to feel the heartbeat of the Earth." Harriet Lawrence Hemenway and Minna B. Hall would be proud. The mothers of conservation are alive and well.

But even as González and Carbajal build a new generation of stewards for the birds and beaches of San Antonio, and as they inspire so many young women to follow in their footsteps, knots have declined there, from 20,000 in 1996 to 2,000 in 2014. Ninety percent are gone. At the moment the cause is not fully clear. Their supply of food may be thinning for a reason still poorly understood. Piletas blasted into the restinga, ATVs, and people still crowding the beaches may be eroding this home. As another rung on the knot ladder to the Arctic weakens, González and Carbajal and Inalafquen may be called to summon all the goodwill and all the allies they have built up over the years, and all the persistence and courage they brought to bear on the soda ash factory. They succeeded then; they may do so again. The sea is still nonnegotiable.

Saint-Exupéry, as the first director of Argentina's airmail service, created and flew the initial routes in the late 1920s, including a route in Patagonia. In the elegiac *Night Flight,* where Fabien pilots a flight destined for San Antonio Oeste that never arrives, Saint-Exupéry evokes Fabien's longing to be aloft, the beauty and spiritual peace he feels flying through a starry night sky as well as the solitude, freedom, and danger. His flight, like that of the birds, holds risk. As the novella opens, night approaches. Fabien lands in the Patagonian town of San Julián. Ten minutes later, he takes off for the sleepy village of Comodoro Rivadavia. Fabien is breaking in the new 1,500-mile run from southern Patagonia to Buenos Aires, stopping every few hours to pick up the mail. He leaves Comodoro Rivadavia at ease, "snugly ensconced" in a sky studded with stars, in a night where he finds "a vast anchorage, an immensity of blessedness," and where he falls "into the deeply meditative mood of flight, mellow with inexplicable hopes."

Fabien navigates with only a compass and gyroscope, and weather conditions telegraphed from airports along the way. The perils of the night, if they appear, can't always be anticipated. He says they are like "worms in a fruit; a fine night, but they would ruin it." Without warning, the calm, clear skies give way. Heavy clouds snuff out the stars, and a gale, rushing in from the mountains, assails him. Winds rage at every airport, the telegraph lines suddenly cut. He has an hour and twenty minutes of fuel and no place to land. Fighting to level the chattering plane in the turbulence, he looks up and sees, in the break of the clouds, "like a fatal lure within a deep abyss, a star or two. Only too well he knew them for a trap. . . . But such was his lust for light that he began to climb." High above the storm, he floats free in the dazzling light, "rich beyond all dreams, but doomed." No one knows where the plane goes down.

On another night in San Antonio Oeste, as the light fades and night falls, another flight readies to depart. A front is coming through, and the wind is turning, now blowing from the south. At the port beaches the birds are restless, jittery. They take off, alight, and take off again. Each time the flock grows larger. Then, with a swooshing sound of air beneath their beating wings, they suddenly lift off, heading north, rising higher and higher in the sky until they disappear. Like Fabien, they will fly through the stars, leaving the Southern Cross behind, staying with Orion, whose bright belt and sword the Argentines call Las Tres Marias, the Three Marys, and *el puñal,* the gaucho's dagger. Flying at 40 miles an hour, they are heading for Brazil, and beyond.

Four

BAY OF PLENTY

Delaware Bay

T he boat scrapes against the beach. Masses of spawning horseshoe crabs pile up along the waterline. Thick crowds of shorebirds scurry along the sand, grabbing loose crab eggs. The day is
overcast and rainy. Neither man nor bird nor crab seems to care.
The men unload beach chairs and spotting scopes: Nigel Clark from the
British Trust for Ornithology, who after so many springs on this bay now
considers Delaware a second home; Richard du Feu, network engineer from
the University of Lancaster, who comes from the freshwater of England's
Lake District; and Bram Verheijen, a researcher from the Netherlands. We
set up the chairs, adjust the scopes, and begin searching for flagged birds.
There are plenty. In Bahía Lomas or San Antonio Oeste, shorebirds would
have abruptly taken flight if we'd arrived this noisily. Here, the beach seethes
with birds. Intent on eating, they rush across the sand, oblivious to us.

Horseshoe crabs plow through the wet sand like armored tanks. It's
hardly an orderly invasion. Desperate to spawn, they clamber over each
other, climb my Wellies, and wedge themselves beneath my tripod, their
barnacle-laden shells clicking against each other. Some are the size of dinner
plates. Bahía Lomas was empty, quiet, and peaceful. Here, the birds are
making a din. The congestion is like Grand Central Terminal at rush hour.
A peregrine swoops. En masse the birds rise. The raptor dives at a knot. The

knot panics, flying into du Feu's scope. The peregrine leaves; the flock resettles. We're on Back Beach, a thin spit of sand in Mispillion Harbor, Delaware. Shorebirds, exhausted and famished after the long flight from South America, arrive in Delaware Bay, one of the nation's largest and least-known estuaries. Ignored by millions of people flocking to Delaware's ocean beaches and the Jersey shore, Delaware Bay, poor sister to better-known Chesapeake Bay, has long held one of the country's best-kept secrets. For a few weeks every year, as human crowds head to the sea, this bay's beaches are awash in swarming horseshoe crabs and ravenous shorebirds—the world's largest concentration of spawning horseshoe crabs and the largest and most important stopover for shorebirds on the eastern seaboard of the United States. For years, so it seemed, only the locals knew.

For most of the year, the large, ungainly, fossil-like crabs live offshore in the deeper waters of the continental shelf, but each spring, on the rising tides of May's new and full moons, they come ashore. In exquisite and mysterious synchrony, birds and crabs arrive together: crabs to spawn, birds to feast on horseshoe crab eggs. Timing is key. To reach their Arctic nesting grounds and successfully breed, knots double their weight during their two-week stay here. Easily digested horseshoe crab eggs offer fast, energy-rich food to fuel this leg of their journey. In 1986, ornithologist J. P. Myers described the sight as "sex and gluttony on Delaware Bay." The description fits. Horseshoe crabs, eggs, and shorebirds blanket Back Beach, packed in so tightly the seething mass conceals the sand. The wide reach of Bahía Lomas and the San Antonio port beaches call for a patient, painstakingly slow approach to birds. Here we simply wait as thousands race helter-skelter across the sand, flags coming into view and then vanishing, a sea of knot legs suddenly obscured by smaller, semipalmated sandpipers. The researchers jot down flag numbers, rapidly filling waterproof notebooks. In England's Wash estuary, du Feu scans for knots in distant mudflats. Here he vacations, simply sitting amid the commotion, where we don't even need binoculars to take in the fine breeding plumage of dunlin, each with a striking black patch on its belly, or the distinctive feeding style of dowitchers, whose beaks work the sand like sewing machine needles moving along a piece of fabric—all only a few feet away.

The busy birds probe the sand or peck at loose eggs at the surface. It's raining, it's windy, and we're wet, but it's warm, and even though we've

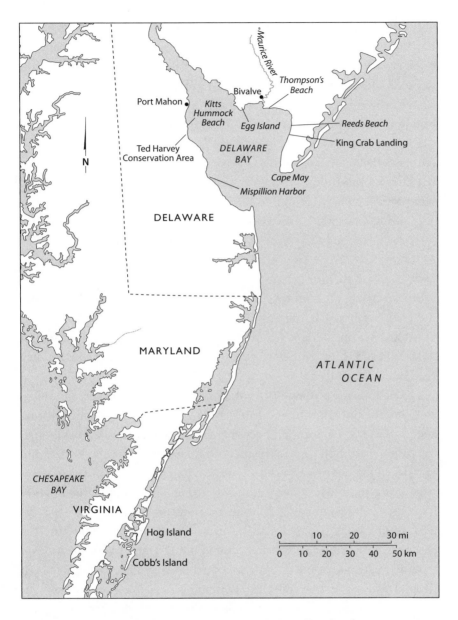

Delaware Bay and Virginia (map by Bill Nelson).

been here for six or seven hours, we can't bear to leave. The tide is ebbing, but the birds haven't departed. Clark, sustained by his enthusiasm, has eaten only a handful of M&Ms and almonds all day. Mesmerized by the scene unfolding before me, I haven't noticed time rushing by either. Kevin Kalosz, who runs the Delaware shorebird program, calls a few times, gently urging us to return. The sun is setting and guests are coming for dinner. Clark won't—can't—pull himself away. So many birds, so many flags. He estimates that on this tiny piece of beach, we're sitting in the midst of 4,000 red knots, 5,000 ruddy turnstones, 5,000 dowitchers, 5,000 semipalmated sandpipers, and 15,000 dunlin—a shorebird mecca.

Knots from South America have flown as many as 7,500 miles to reach this beach, working their way up the coast from Tierra del Fuego and San Antonio Oeste. Some stop in the lagoons of southern Brazil's Lagoa de Peixe National Park, or the sandy mudflats and mangroves of Maranhão in northern Brazil. One knot, YoY, flew across the Amazon rain forest from the Uruguay-Brazil border to Ocracoke, North Carolina, a 5,000-mile nonstop flight in six days. Another, 1VL, left northern Brazil on a 4,000-mile nonstop flight over the Atlantic, arriving in Delaware Bay six days later. No wonder the birds are hungry.

Cape May, a birding hot spot, has attracted outstanding naturalists and birders since the nineteenth century, including the Smithsonian's Spencer Fullerton Baird and Roger Tory Peterson, yet the wider ornithological community didn't "discover" the massive feeding frenzy in the bay until 1977, when Jim Seibert, a local decoy carver, and his wife, Joan, called naturalist Pete Dunne, who'd just begun working at New Jersey Audubon's new Cape May Observatory. As Dunne recounts in *Bayshore Summer*, his portrait of the rhythms of life in Delaware Bay, the Seiberts told him "the beach in front of their house was awash in birds." Dunne, who has gone on to write many books and essays about birds and birding, found the beach covered with horseshoe crabs and semipalmated sandpipers, sanderlings, ruddy turnstones, and knots—more knots, he writes, "than were estimated to be in all of North America." He hadn't seen anything yet. In May of 1981 and 1982, Dunne, Clay Sutton (whose *Birds and Birding at Cape May*, authored with his wife, Pat, led me to birding beaches, marshes, and trails I would have missed otherwise), Wade Wander, (an expert in counting birds), and renowned ornithologist David Sibley (whose field guide, along with several others I carried everywhere I traveled in North America) made an aerial survey of this avian Serengeti. Once aloft, these seasoned, experienced birders, like Morrison and Ross in South America, couldn't believe what they were seeing.

Dunne, who writes that he is "no stranger to numbers of birds" and who has seen flocks of swallows blocking the sun and snow geese rising "like storm clouds," was dumbfounded to find 420,000 shorebirds, 95,000 of them red knots, blanketing the beaches—just a portion of the shorebirds migrating through the bay each spring. That estimate came in at 150,000 knots out of 1,500,000 shorebirds, making Delaware Bay one of the nation's most important spring stopovers for migrating shorebirds. In eastern North America, writes J. P. Myers, who did his early research on sanderlings, "no other spot comes close." In a country where much of the landscape is known or at least traversed, we missed the very heart of this migration. Why?

The written record of human history is, like the fossil record, incomplete. Not everything is written, and much that is disappears. Dr. Maurice Beesley, in his "Sketch of the Early History of the County of Cape May," wrote in 1857: "Isolated as it was . . . and with a sparse population, we find no material to consult, except a meager court record; hence the inquirer is compelled to seek from musty manuscripts and books in other places, a goodly portion of the little that has escaped oblivion. . . . and that little must necessarily be made up of scraps and fragments." The history of shorebirds devouring horseshoe crab eggs on Delaware Bay—at least the history that has been uncovered so far—is meager. Examining it, trying to find more, was frustrating, addicting, and revealing, both for what was there and what wasn't—and why.

In the early nineteenth century, New Jersey was a destination for two men who would become two of the country's first preeminent ornithologists. John James Audubon traveled to the Jersey shore, taking an overnight ride to Egg Harbor, near Atlantic City, with a fisherman-gunner in a Jersey wagon "laden with fish and fowls." He lodged at the water's edge, ate oysters, fished, and searched for heron, marsh hens, terns, and fish hawks (osprey). He made that trip in June. In May of 1829, during another visit, he noticed at nearby Great Egg Harbor an "immense number" of laughing gulls rising from their nests at sunrise, flying west toward the Delaware River, and then returning at sunset. He was certainly in New Jersey at the right time for

Clockwise from bottom: ruddy turnstone, red knot, and two dunlin (drawing by Michael DiGiorgio).

shorebirds and horseshoe crabs, but possibly in the wrong place. Something big was clearly going on over in the bay, but in his account of this trip, we are left hanging about what it might be.

Alexander Wilson, another of America's early great ornithologists, traveled six times to Great Egg Harbor between 1810 and 1813. He lodged at Thomas Beesley's tavern at the ocean's edge. He made his way at least once to the bay's quieter waters but I found little recorded about these bay sojourns. In his quest for birds, Wilson walked almost all the way from Philadelphia to Niagara Falls, sometimes covering more than 40 miles a day. He rowed a skiff 720 miles down the Ohio River in February, just as the ice was breaking up. He slogged with his horse through the dark and dank swamps of the Natchez Trace. Perhaps he didn't write to friends or family about shorter trips whose challenges may have seemed dull in comparison. His journal, where the descriptions may reside, disappeared long ago: his friend George Ord, sportsman and naturalist from Philadelphia, was the last to quote it, in 1828.

Wilson's American Ornithology suggests he knew where Audubon's laughing gulls were going, and why. Wilson saw, where Fishing Creek empties into Delaware Bay, "great multitudes" of laughing gulls feeding upon "the remains of the king [horseshoe] crabs"—so many that when they rose in flight, the sound of their squawking carried for two or three miles. In my lifetime, I will never hear that many birds. Wilson wrote that farther up into the bay, at the mouth of the Maurice River near Egg Island, he saw "bushels" of horseshoe crab eggs "lying in hollows and eddies . . . while the Snipes and Sandpipers, particularly the Turnstone, are hovering about, feasting on the delicious fare." During May and June, he reports, the turnstone lives "almost wholly on the eggs, or spawn, of the great king crab, called here by the common people the horse-foot." Which snipes did he see? Which sandpipers? He doesn't say.

When he does write about knots, he describes them in gray, nonbreeding plumage—his ash-colored sandpipers—feeding on tiny shell-fish the size of apple seeds in the late summer and fall, as were all the shore-birds he saw then. The horseshoe crabs have already returned to the sea. Of the knot in breeding plumage—his red-breasted sandpiper—he doesn't mention where he saw it, so we can't tell whether horseshoe crab eggs were a possible fare.

I read most of Wilson's letters in reprinted collections, then looked at originals held in the Ernst Mayr Library at Harvard's Museum of Comparative Zoology. Written over 200 years ago, their paper is frayed and stained, the ink faded. Their content holds no surprises, but their fragility speaks of all that Wilson—and others—might have observed in Delaware's quiet bay but never put to paper, and all that might have been recorded but then rotted away.

J. P. Hand, a resident of Cape May County, local historian, and decoy carver, generously looked for more scraps of this history and sent me a large folder of articles from old newspapers. He'd found an 1853 report from the *Philadelphia Inquirer*'s correspondent in Cape May. The reporter, befriending an "Old Salt" on a rainy day when he had nothing to do, asked him about clouds of mosquitoes, disappearing Blue Point oysters, and "the utility of king crabs." He was immediately informed that horseshoe crabs "feed seabirds in the spring with their eggs." I wish I'd been sitting in on that conversation. Did the Old Salt see only gulls? Or shorebirds as well?

There's not much history of knots devouring horseshoe crab eggs in Delaware Bay in the late nineteenth and twentieth centuries, even though the nation's best-known ornithologists came to New Jersey. Masses of migrating shorebirds descend on Delaware Bay for a few weeks and then they are gone. It's entirely possible that visiting ornithologists and sportsmen missed this brief window. Witmer Stone, author of the two-volume *Bird Studies at Old Cape May,* summered in Cape May, on the ocean, during July and August—too late for the pulse of shorebirds and horseshoe crabs.

He records hundreds, possibly thousands, of shorebird sightings, listing every beach where his colleagues and he saw birds. Over and over, the same names appear: Two Mile Beach, Five Mile Beach, Jarvis Sound, Grassy Sound, Hereford Inlet, Stone Harbor, Seven Mile Beach, Townsends Inlet, Ludlams Beach, Corsons Inlet, Pecks Beach, Great Egg Harbor Inlet, Absecon Inlet, Brigantine, Little Egg Harbor, Long Beach, Barnegat Bay, Point Pleasant. Every beach faced the ocean. He looked to the east for birds. Horseshoe crabs were west, in the bay. Naturalists who studied fish—curator of fish at Philadelphia's Academy of Natural Sciences Henry W. Fowler and Richard Rathbun from the U.S. Fish Commission—found horseshoe crabs were relatively rare along the Jersey shore: the animals preferred the more tranquil waters of Delaware Bay, where they were unusually abundant.

I wanted to read a natural history of shorebirds in the bay: Stone's shorebird history was of the beach and the adjacent marsh. Charles Urner undertook a 10-year study of shorebirds migrating along the New Jersey coast. "Coast" in his study meant the ocean. Despite the dearth of information, a few secondhand reports allude to what would later be understood as the primary engine fueling spring shorebird migration here. Walker Hand was a lifelong resident of Cape May, high-ranking employee of the postal service, and avid sportsman. Like Stone's other contributors, he doesn't report from the bay, but on the subject of ruddy turnstones, he "quotes an old tradition of a regular flight back and forth across the Cape May Peninsula to feed on king crabs which are washed up in numbers on the Bay shore." Banker and ornithologist Julian K. Potter does the same in a birding magazine. "Thousands of shore-birds were also reported from the Delaware Bay shore on the 20th [of May, 1934]," he writes, and "according to native reports, they were attracted by king crab spawn on which they were feeding." I'd like to see more reports from these "natives."

I found a direct observation from the bay. Harold N. Gibbs, shellfish expert, waterfowl carver, sports fisherman, and administrator of fish and game for Rhode Island, visited Delaware Bay in May 1948. He saw shorebirds, including knots, black-bellied and semipalmated plovers, dowitchers, and other sandpipers feeding on horseshoe crab eggs.

These observations are but fragments, shining light into a past beyond the reach of memory, and otherwise obscured. They hint that Delaware Bay shorebird dependence on horseshoe crab eggs may be at least as old as the bay's recorded history. Did others, visiting or living on the bay, write of the migration in their diaries or correspondence? Or, like Boris Cvitanic, watching the annual migration from his estancia in Bahía Lomas, did they not share their observations with the wider world? I imagine there are reports I haven't found, but so far, each lead has petered out.

An internationally renowned authority on oysters from Rutgers, Thurlow C. Nelson, lived and worked on Delaware Bay in the summer. In 1939, his sister, Theodora, completed her Ph.D. on the spotted sandpiper, one sandpiper easy to identify (as it walks, its tail bobs). A professor at Hunter, she was one of the only paid female ornithologists of her time. Surely if she'd seen the migration, she would have recognized its significance. I wondered whether she had visited her brother at his cottage on

Delaware Bay, and what she might have observed. I can't find her papers, if she left any, though librarians across the country helped me search. I spoke with one of her students who, inspired by Nelson, had gone on to become a respected and well-known shark biologist. She's now 90 years old. Much as she wanted to, she couldn't help me, and neither could another student who'd also stayed in touch with their beloved teacher over the years. The Cape May County Library has an archive of at least 10 local newspapers dating from 1859. They haven't been digitized and they lack indices. I plan to return to Delaware Bay during the gray, gloomy days of winter and read them. We may never know the number of shorebirds passing through Delaware Bay in Wilson's time, yet I keep chasing the history, hoping to find pages that may be missing.

The numbers of horseshoe crabs then may have far exceeded anything I have known or seen. In 1857, horseshoe crab eggs on Delaware Bay were "so thick" they could be "shovelled up and collected by the wagon load." Of the horseshoe crabs that laid those eggs, Wilson reported that at the mouth of the Maurice River near Egg Island "their dead bodies cover the shore in heaps, and in such numbers, that for ten miles one might walk on them without touching the ground." Accustomed to scarcity, we can barely comprehend abundance of this magnitude. Today I couldn't walk 10 feet, let alone 10 miles, on the backs of dead horseshoe crabs. It's tempting to think Wilson may have been referring to live animals, but the construction of the sentence is clear, and Wilson, who identified 26 new species of birds, corresponded with Thomas Jefferson about nuances in crest and color among jays, and is honored by more species bearing his name than any other American ornithologist, presumably could distinguish the dead from the living, leaving us to face the measure of our losses.

They are substantial. The waters of Delaware Bay once teemed with life. It is worth briefly considering this past as a frame for today's difficult struggle over the bay's restoration. In the late 1800s, thousands of sturgeon swam up the Delaware River, ripping through and destroying the nets of shad fishermen. Jumping sturgeon nearly capsized their boats, even occasionally landing inside. More Atlantic sturgeon—some 180,000 fish—spawned in the upper reaches of Delaware Bay than anywhere else in the United States. The fish were enormous—14 feet long and weighing as much as 800 pounds.

At Benny's Buoy and Caviar Point, fisherman caught roe-laden sturgeon. They smoked the flesh and mixed the roe with high-quality German salt, drained the brine, and packed the roe (now caviar) into wooden casks. By 1888, Delaware Bay, the nation's capital of caviar, yielded 75 percent of the U.S. sturgeon catch; trains left Caviar Point every day with full casks bound for New York, Philadelphia, Europe, and even Russia.

Dusky, thresher, hammerhead, and sand sharks, the latter almost as big as sturgeon, swam in the bay. In 1856, fishermen caught 500 sharks at the mouth of Fishing Creek, extracted oil from their livers, and sold the bodies as compost.

Each spring, millions of shad ran up the Delaware, swimming toward Trenton to spawn in the streams of their birth. In 1897, Marshall McDonald, writing for the U.S. Fish Commission, found fresh shad "in the height of their season, one of the most delicious of the finny race." He recommended either "planking" them—nailing them to a clean oak plank and roasting them on a brisk fire—or smoking them in a rub of salt, saltpeter, and molasses. The fish were large and heavy. Wilson reported that after an osprey near Beesley's tavern finished picking at a shad, the remains weighed six pounds. In the 1890s, more shad—19 million pounds—were fished from the Delaware every year than from any other river along the Atlantic.

Oysters carpeted the bay. Delicious and sweet, they too commanded a large market. Schooners lined the wharves in Bivalve, at the mouth of the Maurice River. Warehouses, chandleries, and a customs office lined the streets. In 1892, 4,300 people worked in the New Jersey oyster industry. That year, oystermen took more than 1 million bushels of oysters from the bay, either by dredging or tonging. Shuckers reported that some were as large as their hands. In 1886, 90 boxcars filled with oysters left Bivalve each week for northern markets. As recently as 1964, shuckers, many of them African American, still lived in nearby Shell Pile, in dilapidated row houses lacking running water and central heat.

We depleted this plenty, and the horseshoe crabs and shorebirds as well. One of the strongest arguments for the historical dependence of long-distance migrating shorebirds on these ancient mariners from the sea may be that when horseshoe crabs were removed from Delaware Bay, shorebirds, unable to find the sustenance they needed, disappeared as well. The late 1800s saw a near extirpation of horseshoe crabs from Delaware Bay. Horseshoe crabs had been sought, early and often, to feed hungry pigs. At the mouth of the Maurice River, Wilson observed hogs "driven down, every spring, to feed on [king crabs], which they do with great avidity; though by this kind of food their flesh acquires a strong, disagreeable, fishy taste."

Feeding hungry pigs and fertilizing corn rows with horseshoe crabs was not enough to do them in. The end began when Thomas Beesley, of Beesley's tavern in Egg Harbor, figured out how to make money from horseshoe crabs. The enterprising state senator once cooked dinner for an entire family from a flounder dropped by an osprey, and concocted an osprey "egg-nogg." (The egg seemed "perfectly fresh," but the cordial "smelt abominably.") In 1855, Beesley, his attention drawn "to the probable value of the king crab," purchased two miles of bay beach "at a really nominal value" and opened a factory to manufacture fertilizer. He drew up a price list and was soon in business. "From half a mile of beach no less than 750,000 [king crab] were taken . . . the multiplication table ran out before he had counted a few days." The crabs were speared, taken by boat to the factory, "heaped up by the thousands like bricks," and when they died, thrown into a threshing machine, roasted like coffee beans, and finally "passed through a mill and ground into powder as fine as flour." The resulting fertilizer, cancerine, sold for $30 a ton to Maryland fruit growers at half the price of Peruvian guano and, being less soluble and volatile, lasted longer. So began what would become a colossal and indiscriminate removal of horseshoe crabs from Delaware Bay.

Untold numbers of horseshoe crabs were turned into fertilizer. In 1857 on Town Bank, a bay beach close to Cape May, more than 1 million horseshoe crabs were removed along a one-mile stretch. In 1880, more than 4 million were taken from all of Delaware Bay, either by hand in Delaware, or in New Jersey from weirs and nets set along the shore. Word got around that horseshoe crabs were a good bet. Brooklyn racing aficionados disembarking from the ferry onto special trains going directly to the New Jersey Jockey Club's Elizabeth track bet on a favorite and consistent winner—King Crab. I don't know what happened to the horse, but the heavy extraction of horseshoe crabs from the bay couldn't last.

In 1884, Richard Rathbun wrote that horseshoe crabs were greatly reduced from their use as fertilizer and that "a few years more of indiscriminate capture would result in their being entirely exterminated from the region." Hugh McCormick Smith, also of the U.S. Fish Commission, found that "the diminution in the abundance of the crabs is no doubt chiefly due to the unfortunate practice of capturing them during the spawning season, usually before the eggs are deposited or impregnated. It seems probable that before long the decimation will become so pronounced that the profitable prosecution of the fishery will be impossible." The end didn't come that quickly, or with such finality. The crabs held on longer than either man thought possible.

My New Jersey gazetteer showed a place called King Crab Landing. I tried to find it. The old road in had grown over, leaving only a faint trace in the tall grass. The local librarians didn't know the place, but their parents did. The last of Delaware Bay's horseshoe crab fertilizer factories once stood at the end of the road. It was owned by Joseph Camp and operated by his son Franklin during the 1930s. Betsy Haskin, who tends her father's oyster garden in the water near the site of the old factory, put me in touch with the owner's great-grandson, Barry Camp. Franklin Camp's son Willets Corson Camp remembers his grandfather's mile-long pound net collecting not only crabs but "every kind of fish in the bay." At low tide, his father "pitchforked" crabs from the pound into a scow that held about 3,000 crabs. A towboat named *Rescue* brought the scow to a pier where a conveyor belt loaded crab into a small railroad. A gasoline-powered engine towed the car to the factory. There, horseshoe crabs were crushed, fed into one of three cookers, and steamed. Fresh crab meal, spread out in a layer about two inches thick, dried in a long shed nearby. Occasionally the crab meal caught on fire. Willets Corson Camp helped in the factory, filling 100-pound bags of crab meal and dragging them off the scale. The bags were sold to the fertilizer distributor and manufacturer I. P. Thomas & Son in Camden. Willets's sister, Frances Camp (now Hansen), still remembers climbing on bags of fertilizer piled high in the factory and the "terrible odor" of dying crabs.

Fran Camp had a childhood friend, Marjory Nelson (Thurlow Nelson's daughter and Theodora Nelson's niece), who lived on the bay in the summers. The family's summer cottage, Limulus Lodge, named after the horseshoe crab, *Limulus polyphemus,* was near the Camps' railroad track. Marjory Nelson, recalling her childhood on the bay, offered another reason why Philadelphia's naturalists looking for birds headed for New

Jersey's ocean beaches. "It was just too hard to be here," she says. Biting insects swarmed unpleasantly. "Our cottage was built in a swamp. I remember my mother standing in clouds of mosquitoes hanging out the wash. On the way to the privy, we'd take the Flit gun [an insecticide popularized by ads drawn by Dr. Seuss] and spray it ahead of us as we walked." Cape May was a resort, with comfortable rooming houses and ocean breezes. The bay "was a working-class community where people struggled to make a living. It reeked of rotting crabs and fertilizer. My mother burned incense to cover the smell." Willets Corson Camp remembers the smell, too. "When didn't it smell? It smelled constantly, a stench at all phases."

When Joseph Camp died, the factory closed, the land was sold, and the family moved into town. Willets Corson Camp is getting on in years. His son helped piece together his father's memories. "We were the bad guys," Barry Camp told me when I first inquired about the factory. I'm not so sure. Back then, no one really understood the implications of emptying the bay of horseshoe crabs. And, as would become apparent later, the history of horseshoe crabs is a history not only of decimation but also of recovery.

Delaware Bay was emptied of horseshoe crabs. Outside the bay, in places where people weren't turning hundreds of thousands of horseshoe crabs into fertilizer, were shorebirds eating their eggs? They were. Up in Massachusetts, Warren Hapgood, sportsman, well-respected authority on shorebird shooting, and frequent contributor to *Forest and Stream,* wrote in 1881 that red knots, which hunters variously call robin snipe, grayback, knot, and red-breasted sandpiper, "have a *penchant* for 'horsefoot' eggs, and display considerable ingenuity" in ferreting them out—scratching the sand and "poking out the eggs with their bills." In 1912, knots were still eating horseshoe crab eggs in Massachusetts. State ornithologist Edward Howe Forbush wrote then that knots "are fond of the spawn of the horsefoot crab, which, often in the company of the Turnstone, they dig out of the sand." Both men remarked that knots fight with ruddy turnstones over horseshoe crab eggs.

In 1940, Charles C. Sperry, specialist in the eating habits of birds at the U.S. Department of the Interior, published what appears to be the very first scientific data documenting the consumption of horseshoe crab eggs by red knots. His study analyzed the stomach contents of knot, dowitcher, snipe, and woodcock. The knots were collected between 1911 and 1918 from May through September along the Atlantic and Gulf of Mexico coasts as well as in Ontario, Canada, and Greenland. In the spring, both knots and dowitchers were eating horseshoe crab eggs: one knot's stomach was filled with nothing but horseshoe crab eggs. A dowitcher's stomach held 300 eggs. Michael Haramis, research biologist with the U.S. Geological Survey's Patuxent Wildlife Research Center, had been helping me collect historical observations of knots eating horseshoe crab eggs. He examined Sperry's old, handwritten note cards, still on file in the archives at Patuxent. Knots in Sperry's study came from Alabama, Florida, and South Carolina. Not one came from Delaware Bay. "The men who collected these birds," Haramis wrote me, "went to places they liked," and "in those days they took what they could get. Apparently Delaware Bay was not a popular place for them and they had no immediate connections there." Once again, Delaware Bay was overlooked.

In all likelihood, shorebirds disappeared from Delaware Bay when horseshoe crabs were fished out. Neither Dunne nor Sutton found any history of shorebird gunning on the New Jersey side of the bay then—no decoys, hunting clubs, hunting records. Shorebirds, Sutton writes, "simply were not there. If they had been, they would have been hunted." At the same time as millions of horseshoe crabs were being turned into fertilizer on the bay, elsewhere thousands, perhaps hundreds of thousands of shorebirds were being shot—for science, for sport, and for food. Ornithology was built on a foundation of dead birds. Distinctions among species and subspecies; differences among fledglings, juveniles, immature and mature adults, and males and females; diet: scientists learned much of this information by shooting birds. Audubon killed tens of thousands, then resurrected them in his paintings. Darwin's ideas of evolution emerged from fossils he studied and birds he shot.

As a youth, Darwin loved killing birds, and fondly remembers killing his first snipe, a shorebird. "In the latter part of my school life I became passionately fond of shooting, and I do not believe that anyone could have shown more zeal for the most holy cause than I did for shooting birds. How well I remember killing my first snipe, and my excitement was so great that I had much difficulty in reloading my gun from the trembling of my hands.

This taste long continued and I became a very good shot." In Cambridge, he practiced in his room, firing at a lighted candle waved by friend. He didn't let loose any gunpowder: if his aim were true, a puff of air shot from the barrel extinguished the light. The sound of the exploding cap led the college tutor to think Darwin liked to spend hours in his room cracking a horsewhip.

For many naturalists, the urge to collect was irresistible. Walker Hand and a friend, making the first recorded sighting of a parasitic jaeger in New Jersey in July 1922, immediately tried to kill it, "but as they had no firearms, it flew off in the bay unharmed." During his year studying birds on the Arctic's Southampton Island, ornithologist George Miksch Sutton, whose work I'd come to know when I traveled there in search of knots, shot or tried to shoot at least one, and often more, of almost every species of bird he saw: females with eggs; downy young; the "rare" Iceland gull and the "uncommon" Hudsonian curlew; and five knots. He could skin, stuff, and sew up a bird in 15–20 minutes, "with every feather in place"—at least 17,500 birds over the course of his career.

Sportsmen loved killing shorebirds. "There are few more exciting experiences in the sportsman's life," wrote Robert B. Roosevelt, sportsman, Democrat, and uncle to Theodore, "than 'whistling up' a flock of bay snipe to the decoys." In one account of spring shooting for robin snipe—the hunters' name for knots—hunters arrived at the Jersey shore at the end of May to find "countless numbers of robin snipe and bullhead plover." They "artistically" arranged decoys to simulate a flock of feeding birds. Then, hidden in piles of seaweed, staying dry on India rubber blankets, they used "nothing more than a school boy's penny whistle" to call in the knots as they came in on the rising tide. "Flock after flock of robin snipe were whistled up to our decoys, and great havoc was made each time in their ranks." So much havoc that by the time Witmer Stone began regularly visiting Cape May, most of the horseshoe crabs had been taken for fertilizer and many larger shorebirds had been shot as well.

"Everybody shot the knot, both fall and spring," state ornithologist Edward Forbush wrote, "for it was in demand for the table, brought a good price in the market, decoyed easily and flew in such flocks that many could be killed at a shot." Although some found knots "only fair eating, being a little fishy in flavor," people ate them for at least 450 years. The bill of fare at an Oxford University commencement dinner in 1452, hosted by George Nevill, brother of the Earl of Warwick, included "plover; knottys; styntis; quayles." Knots, not considered the most elegant fare, were served at the "third table." Diners at the chief table ate "fesant in brase" and "swan with chowder."

Three hundred years later, the knot's culinary status had improved. New York's luxurious Astor House, whose guests included Abraham Lincoln, listed on its October 11, 1849, menu roasted wood ducks, dowitchers, plovers, mallards, and broiled robin snipe. An article in the June 11, 1887, issue of *Good Housekeeping* entitled "Table Supplies and Economics: What to Buy, When to Buy, and How to Buy Wisely and Well" praises the offerings in a New York market, which include robin snipe at $1.75 a dozen, smaller yellowlegs at $1.50, and greater yellowlegs at $3.00. Henry Fleckenstein, author of many books about bird decoys, wrote that birds were "hauled from the meadows in wagons heaped full over the boards," packed in barrels, and shipped by train or boat to city markets. Birds that weren't quite as good to eat were used as packing for the others. Not all the shorebirds made it to market. Barrels of knots, turnstones, and plovers shipped to Boston spoiled during passage and were tossed overboard. Hunting so reduced the knots, sportsman and naturalist George H. Mackay writes, that they "in a great measure have been killed off" and "are in great danger of extinction."

The Lacey and Migratory Bird Treaty Acts brought a halt to market gunning. Naturalists tracking shorebirds had this to say: "The increase in numbers of these birds subsequent to the cessation of shooting is gratifying. However, it cannot be hoped that these birds will ever re-attain their former numbers." Of the knot, this: "Excessive shooting, both in spring and fall, reduced this species to a pitiful remnant of its former numbers . . . it has increased slowly . . . but is far from abundant now." Synthetic fertilizers took the pressure off horseshoe crabs. By spring 1977, 75,000 horseshoe crabs were spawning on a mile-long stretch of beach near Green Creek, hardly the million that had come up on the beach 100 years earlier, but substantially more than the mere hundreds seen in 1951.

I continue to search, perhaps futilely, for more shreds of a history that may have been obliterated. Still, our meager understanding of the immense richness of the sea has been upended before. Scientists delving into the history of endangered green sea turtles discovered that they once filled the Caribbean Ocean, perhaps literally. The 2002 population of 300,000 is but a remnant of the 91 million adults living there in the 1600s. The historical abundance of shorebirds and horseshoe crabs, and whether, where, and how much they depended on horseshoe crab eggs, matters. We so easily settle for the diminished world around us, a world that, in terms of the richness and abundance of plant and animal life, may be a mere 10 percent of what it once was. Unaware of what we have lost, we can't imagine what we might restore, and instead, we argue over how many of the scraps we might still take.

Horseshoe crabs and shorebirds survived a slaughter of gargantuan proportion. How well knots had recovered in Delaware Bay when Dunne, Wader, and Sibley flew over a beach covered in shorebirds isn't known. Their 150,000 knots may be merely a wisp of what had been there previously. What is known is that the frenzy of horseshoe crabs and shorebirds I saw on Mispillion Beach was only a fraction of what they'd seen 30 years earlier.

Five

TENACITY

We piece together the evolution of life on Earth recognizing that very few plants and animals leave a record of their time here. Of those with hard shells or bone that can withstand the ravages of time, very few—only 2 to 13 percent—ever become fossils. If knots left a fossil trail of their evolution, it is still to be found. Scientists mapping the genes of sandpipers believe that knots came into their own 11 million to 16 million years ago, when the planet was warmer and drier, when alligators were living as far north as England, when the great apes that would give rise to humans had yet to walk upright. Knots diverged, evolutionarily speaking, from their closest relative, surfbirds, stocky sandpipers that today live along the West Coast of the United States.

From then until the last Ice Age, their story is untold. In her genetic reconstruction, set against the backdrop of Earth's climate history, Deborah Buehler considers how ancestral knots may have divided into the six lineages we know today—all breeding in the Arctic, but migrating to destinations across the globe. As they evolved, they weathered enormous stress. As the ice spread to its greatest extent, 18,000 to 20,000 years ago, polar deserts crept south, and the tundra shrank, dividing and isolating their nesting grounds. One population became two. Today, *Calidris canutus canutus*, the oldest knot lineage, nests on Siberia's Taimyr Peninsula and winters in western Africa, on mudflats built by Saharan sand. The other population, nesting closer to the Bering Strait, flew east in a warm spell, crossing the water into North America. A final cold snap divided them—one group on each side of the Bering Strait—and two populations became three.

Three became four when the glaciers receded, larches and birches advanced to the ice edge, and the tundra shrank again, this time dividing knots on the Russian side of the Bering Strait. Both lineages migrate through the Yellow Sea. One, *Calidris canutus piersmai*—named after Theunis Piersma, who studies the mind-bending physiological adjustments of knots on their long migrations—winters in northern Australia's Roebuck Bay, while the other continues on to southern Australia and New Zealand. The Earth continued to warm, forests advanced north, farther than they are today, and divided North American knots. Four populations became today's six. *C. c. roselaari*, perhaps the smallest population of them all, breeds in Alaska and the Wrangel Islands and migrates along the Pacific. *C. c. islandica* left the Atlantic flyway, taking a 600-mile shortcut through Iceland and Europe. Perhaps a storm blew them east, or perhaps a few young birds, unsure of their route, accidentally veered east to Iceland and then flew on to winter in Europe's Wadden Sea. I'm following *C. c. rufa* up the Atlantic coast.

The climatic convulsions that ended the last Ice Age squeezed knots out of their homes not once but several times. During these upheavals, the population crashed, dropping to as few as 500 female birds. They survived. Perhaps they can, and will, continue to survive as the Earth warms again, but they face different and potentially more complex pressures now. Today, the lives of red knots migrating north from Tierra del Fuego are intertwined with the lives of an ancient animal, one that lives in the sea and came into being long before the very first bird took flight.

Almost all species, 99.9 percent, that ever lived on Earth are now extinct. The record of their existence, preserved in fossils, tells us who rose to dominate continent and sea, who lived quietly in the shadows, who disappeared, and who endured. Horseshoe crabs, animals named for the horseshoe shape of their shells, more closely related to spiders and scorpions than crabs, endured. They outlasted fellow inhabitants of the ocean now long gone, and even the ocean basins themselves, living on when ancient seas drained away, their basins raised into the mountains of continents.

Red knots and horseshoe crabs on Delaware Bay (drawing by Michael DiGiorgio).

Horseshoe crabs persisted, finding their homes in younger, new seas, like the Atlantic.

These antediluvian-looking animals are ancient, but just how ancient came as a surprise. Not every rock that contains lines from Earth's history has been uncovered; the discovery of new fossils still reconstructs our understanding of the past. In 1823, for example, a young Englishwoman, Mary Anning, digging in the fossil-rich cliffs of Lyme Regis, Dorset, stunned the scientific community with her recovery of the skeleton of a plesiosaur, a giant marine reptile whose existence had hitherto escaped their notice. In 2004, after four seasons of picking, scraping, and digging on the Arctic's barren Ellesmere Island, paleontologists examining a rock exposure that extended for 1,000 miles removed yet another rock—and found the snout of tik taalik. A key evolutionary link between fish and amphibians, its fins were evolving into limbs. The discovery of a fossil from a surprisingly old horseshoe crab was more serendipitous. An artist in central Manitoba looking for flat rock on which to paint accidentally came upon a fossil of an ancient sea scorpion.

In 2006, Canadian paleontologist Graham Young returned to the site, a remote jack pine forest near William Lake, five hours north of Winnipeg. There, he and his colleagues uncovered a trove of fossils, more sea scorpions and jellyfish and, the following year, what looked like a small horseshoe crab. Only six weeks later, still digging, this time along a rocky shoreline they thought they knew well, in a cove on the west side of Hudson Bay outside Churchill, Manitoba, Young and paleontologist David Rudkin found another fossil horseshoe crab. Their discoveries, 445 million years old, pushed back the age of the oldest known horseshoe crabs a full 100 million years. The paleontologists named them *Lunataspis aurora*—*luna* for the crescent moon contour of their shieldlike shell and *aurora* because they lived much closer to the dawn of animal life than science had previously understood.

Lunataspis aurora would shortly be relieved of its place as the oldest horseshoe crab. Around the same time Young was at William Lake, across the ocean an international fossil collector and dealer in Erfoud, Morocco, showed Belgian paleontologist Peter Van Roy a lovely specimen of an enigmatic and extinct animal, *Tremaglaspis,* a distant relative of horseshoe crabs. Van Roy was then working on his Ph.D. in the oasis of Tafilalt in eastern Morocco. The fossil originated farther south, in the desert of the

Draa Valley near the town of Zagora, but apart from that, the dealer couldn't, or wouldn't, tell him much. He saw nothing from Zagora for two more years, when the dealer tempted him with another fossil, a stunning, exceptionally preserved specimen from the same layer of shale, the Fezouata. It looked like a garden pill bug. Van Roy immediately refocused his dissertation on the Draa Valley. Still a poor graduate student lacking sufficient funds to rent a car, he persuaded a taxi driver to take him there, three hours to Zagora and then another two down a dirt track that has since deteriorated. It was a fateful day. Since then Van Roy and Mohammed Ou Said Ben Moula, who discovered the first two fossils, have worked together in this 200-square-mile area of rocky desert. So far, they've uncovered 3,000 specimens representing close to 100 types of rarely and exceptionally preserved ancient marine animals, with more to come.

Their findings are nothing short of extraordinary, uncovering a vista into a time almost half a billion years ago when the sea teemed with odd and strangely beautiful animals that existed millions of years after scientists thought they'd become extinct. They found fossils of a stubby-legged worm covered in plates of armor, a three-foot-long giant shrimplike animal, sponges, ancient relatives of clams, snails, and sea urchins, and hundreds of complete fossils of horseshoe crabs. Most of the animals from this time in Earth's history are extinct, their stay long lasting compared to ours but brief compared to the tenacity of horseshoe crabs. For now, Van Roy and Ben Moula's fossils reign as Earth's oldest known horseshoe crabs, pushing the record of horseshoe crabs inhabiting Earth's ocean back by a full 30 million years.

The fossils attest to the endurance and versatility of horseshoe crabs. *Lunataspis aurora* lived in a warm, shallow, tropical ocean, while Van Roy and Ben Moula's horseshoe crabs lived near the South Pole in cold, deep water. One day a roiling cloud of mud and dirt triggered by a distant storm high on the continental shelf swept down and buried them. They turned to stone layered into the seafloor. The ancient ocean basin eventually closed, its foundation lifted into the mountains of Morocco. Over millions of years, wind and rain weathered the mountains, exposing the rocky tomb. Van Roy intends to name the Moroccan crab after Ben Moula, to honor the highly talented and gifted man whose findings are adding a new chapter to Earth's evolutionary history.

In the grand reach of time, our stay on Earth has been but a moment, while the horseshoe crab, one of Earth's oldest living animals, has endured 475 million years. If the history of life on Earth were compressed into a single year, horseshoe crabs would appear in the spring, around the solstice, and the first birds would emerge in autumn. Our genus, *Homo*, appears as the year closes, on December 30. The Earth was a much different place when the first horseshoe crabs were crawling over the seafloor. Water temperatures in tropical seas reached as high as 104 degrees. The atmosphere was thick with carbon dioxide, 15 times more than today, and sea level was 700 feet higher. The sun didn't shine as brightly, but the Earth spun faster, shortening the day to 21 hours and lengthening the year to 417 days. Volcanic eruptions wracked land and sea as continents broke apart and young ocean floor was built. It was a time when the diversity of life in Earth's ocean proliferated.

The first bird appeared during the Jurassic period, 150 million years ago. At that time, warm, shallow lagoons at the edge of the ancient Tethys Sea covered what is now the village of Solnhofen, Bavaria. Long after the sea receded and the lagoons evaporated, stonecutters quarrying limestone for building tiles and fine lithographic plates exposed one of the world's richest troves of fossils, 550 species of plants and animals, shining a light onto life in this distant past. Dwarf crocodiles and small dinosaurs lived on arid islands in the lagoons. Squid, fish, crab, and large ichthyosaurs swam near ancient coral reefs. It was here that Earth's primordial bird, *urvogel*, was buried, and then found.

In 1861, only a little more than a year after Darwin published *On the Origin of Species,* the well-known German paleontologist Hermann von Meyer obtained from a Solnhofen quarry an exquisite imprint of a single feather that, he said, "perfectly agrees with a bird's feather." Indistinguishable from those of modern birds, it belonged to Earth's oldest bird. He named it *Archaeopteryx,* "ancient feather." The quarry quickly yielded one skeleton. Nine more would follow.

Archaeopteryx had a dinosaur's teeth, long tail, and claws, and a bird's "wish bone," wings, and feathers. That first single feather, scientists now know, was black, a covert, covering a bird's wing feathers. Its pigment-producing cells, contributing structure and durability, are identical to those found in the plumage of living birds. *Archaeopteryx* lived onshore, perhaps eating cicadas and crickets or beetles and dragonflies, and perhaps perching in ancient ginkgoes or evergreens. Monsoon winds and rain may have blown this dinosaur bird, not the most capable of fliers, out over the lagoons where, exhausted from fighting the wind, it drowned. Their bodies didn't decompose in the oxygen-poor water and over time were buried in the mud and turned to rock.

Five slabs of rock also yielded the death march of a horseshoe crab and, at the end, the fossil of the crab itself. The stagnant, unusually salty, and airless water of the lagoon was a death trap, asphyxiating almost immediately any animal that fell in or was washed in by rain. Crustaceans and clams still alive when they hit bottom moved no more than a few inches along the sand before they died. Prints of the horseshoe crab's tail, shell, and claws embedded in the rock slabs show that it landed on its back in the soft bottom, righted itself, then made its way through mud and sand, struggling, until it succumbed after crawling 32 feet.

That's a long death track in an airless lagoon, especially when other animals perished so quickly. Horseshoe crabs are hardy. They survived all five of Earth's mass extinctions, living on in a hot, acidic, and carbon dioxide–laden ocean that 250 million years ago snuffed out almost all marine life, and surviving when an asteroid hit the Earth and killed the dinosaurs. Eventually, the lives of ancient horseshoe crabs and young red knots would converge, the well-being of knots depending on the well-being of horseshoe crabs, and then the well-being of both determined by humans dwelling at the edge of the sea.

I am eager to be back out on the beach. On the New Jersey side of the bay, one research team gathers near the breakwater at the end of Reeds Beach. Amanda Dey from New Jersey's Nongame and Endangered Species Program waits with a large crew in a dip in the dunes, back from the beach. Closer to the beach itself, biologist Larry Niles, from New Jersey's Conserve Wildlife Foundation, is hidden in the scrub. He focuses his binoculars on birds beginning to fly in as high tide approaches. Another crew member waits by a firing box set up in the dunes. It's wired to three small cannons attached to a 30- to 40-foot net buried in the sand. Farther down the beach, another man crouches in the wrack line. Niles and he both have radios. After a large flock lands and settles, Niles radios him to "twinkle." He crawls

nearer the birds, gently nudging them up the beach toward the camouflaged net. His pace is measured. If he moves too quickly, the birds will take
off. He's skilled, but the flock disperses, spooked by a peregrine. When the
birds return, they land away from the net. Twinkling begins again.

An hour, often two, may pass before several hundred birds are feeding
near the net. I've attended these catches and waited all morning, only to see,
if the wind picks up or not enough birds fly in, Niles call it off. Today, the
breeze is light, a large flock crowds in the catch zone, and Niles gives the
order to fire. The net releases, billows over the birds, and traps them before
they can escape. Everyone races to the water. Dey and others with considerable experience quickly disentangle the birds. The rest of us place them in
boxes in the shade. Sitting on folding chairs in the sun, each of us with an
assigned task, we measure, weigh, and band the birds as well as draw blood
samples.

Niles, who researched raptors for his Ph.D., sits in one chair, carefully
affixing geolocators to banded birds. The evening before, he worked with
volunteers soldering, sewing, and gluing these light sensors to tiny packs
the birds will carry on their legs. He's quite skilled with a blowtorch.
Weighing less than half an ounce, the data loggers record light levels every
10 minutes. If the birds are retrapped and the data downloaded, researchers
can map what until now has been hard to see—all the stops along their
migration path, and how much time they spend in the air and on the
ground. The British Antarctic Survey developed geolocators to track albatross. Ron Porter, an engineer from Pennsylvania, designs and builds the leg
mounts, starts and calibrates the "geos," as he calls them, analyzes the
retrieved data, and turns them into beautiful maps. Niles and he and their
colleagues have deployed 600 geolocators since 2009.

Researchers working in New Jersey, like those in Delaware, come from
across the world. Clive Minton from Australia, who first developed the
cannon net, is here. Patricia González is out on the bay counting birds with
Alan Baker from the Royal Ontario Museum. Baker, with Deborah Buehler,
has carried out genetic research on different knot populations. Guy
Morrison is here. The researchers, a distinctive species in their own right,
are as itinerant as the birds, migrating with them along the flyway, tracking
numbered flags to see who returns and who disappears, seeking birds with
geolocators. They huddle around Porter's maps as if they are receiving news

from a child who has traveled the world and now come home. Recording flag numbers year after year, they've learned that as many as 90 percent of adult knots can return to their wintering and staging areas each year.

The tone on the beach is collegial and friendly, but everyone is concentrating, working steadily to release birds as quickly as possible. I don't much like banding. I don't trust myself to do it quickly, and the more nervous I am, the more the birds squirm. The pros proceed calmly. I weigh knots—an easy enough task. Holding them in the palm of my hand, I feel their racing hearts calm. After giving their weights to the scribe, I turn toward the open sand and water. Opening my fingers, I feel the lift as each bird takes off. I watch every bird as it flies away.

Delaware Bay is one of the country's most important stopovers for migrating shorebirds, along with Alaska's Copper River delta, the marshes and grasslands of Kansas's Cheyenne Bottoms, and Gray's Harbor in Washington. In 1982, Morrison, recognizing that shorebirds concentrated in extraordinarily large numbers in particular areas, proposed a necklace of protected areas along the flyway. J. P. Myers, first from his base at the Philadelphia Academy of Natural Sciences (where Witmer Stone served for 50 years) and then at the National Audubon Society, and Pete McLain from the New Jersey Department of Fish, Game, and Wildlife, made it happen. In 1986, New Jersey became the first site in the new Western Hemisphere Shorebird Reserve Network (WHSRN), with long stretches of the bay coast protected in wildlife refuges. Myers, a Renaissance man, would go on to coauthor a critically acclaimed book on endocrine disrupters and to found Environmental Health News, a news service that fills a critical gap left by newspapers with dwindling resources. WHSRN has now become an international network of 32 million acres of protected shorebird habitat in 13 countries, including sites in Bahía Lomas, San Antonio Oeste, and many other bays I'd visit along the flyway.

Most knots migrating north along the Atlantic, 50 to 80 percent, refuel in Delaware Bay. Since 1989, their numbers here have dropped by 70 percent, similar to the precipitous decline in Bahía Lomas. Mispillion, as crowded and congested as it seems, is but a vestige of the avian Serengeti Dunne had seen so many years before. Myers reported "extraordinary concentrations" of birds on Delaware Bay beaches 30 years ago—100,000 birds on Reeds Beach, 350,000 on Moore's. Nothing I saw came close to that.

Losing more birds could send them on a critical slide to extinction. When so many knots depend on a single area, their numbers won't save them: if their home begins to deteriorate, what endangers a few endangers all that dwell there. Dey and Niles have worked assiduously to prevent that, Niles since 1985. They will travel from one end of the flyway to the other, going into remote, difficult-to-reach places to find red knots. They will explore every avenue that might explain where and why knots have declined, and they will take all the time necessary, years if need be, to reverse the trends.

One of the greatest challenges for knots is on their home ground. Niles began his career working for the State of New Jersey, helping acquire land to protect shorebirds. Today, long stretches of New Jersey bay beaches and wetlands are protected wildlife refuges. In the spring, the state closes most bay beaches for a few weeks when horseshoe crabs are spawning and shorebirds are feeding. ATVs, dogs, and throngs of bathers frighten the birds, who don't always return and then can't find the food they need. Before shorebirds arrive and after they depart, the beaches are open, but during May and early June, tape is strung across the entrances. Signs explain why. I have to admit that after driving to three closed beaches and wistfully gazing at long stretches of sand I couldn't walk, I was tempted to duck under the tape. Instead, I accompanied a couple of local anglers who, like me, were making their way up the coast looking for a beach. They were hoping to catch mullet for lunch. Longtime residents, they understood and accepted the closures. A 2013 study of compliance at New Jersey beach closures found that most people cooperate with and support them, with cooperation lowest among some joggers and dog walkers, who proceeded onto the beach anyway.

Considering how shorebirds were squeezed out in Río Grande and may be in San Antonio Oeste as well, New Jersey's aggressive policies to give shorebirds and horseshoe crabs space on bay beaches were farsighted. Still, providing peace and space for the birds, while necessary, proved to be insufficient. The number of knots continued to drop. Niles thought he understood why.

The day I join Niles and Dey on Reeds Beach is hot but breezy. The days I cross the bay to join biologist Richard Weber, now retired from the

University of Delaware, are hot and still. He and his crew are collecting horseshoe crab eggs from Delaware beaches. To reach our first beach, in the Ted Harvey Conservation Area, we walk through a stand of marsh, greeted by gnats, ticks, and mosquitoes. Horseflies, I am told, will be arriving later. The tide is falling; horseshoe crabs won't spawn until evening. Before collecting cores of sand that may contain eggs, Weber taps an overturned crab with his foot. It stirs feebly. He flips it. Crabs overturned at the water's edge can eventually right themselves with their long, spiky tails, but if they've stranded high in the sand, they can't maneuver. It's painful to watch a flailing crab, its claws futilely spinning, grow exhausted. On high-energy beaches, as many as 10 percent of spawning crabs strand. This beach is calm. The waves barely ripple. Bird tracks are everywhere. The beach is eroding, exposing patches of peat. Horseshoe crabs, Weber tells me, balk at peat.

It's a warm morning in May, the water temperature is running at 62 or 63 degrees, he says, and "the crabs are enjoying it." At Kitts Hummock Beach, just south of Dover Air Force Base, the tide has ebbed across a mudflat cobblestoned with buried horseshoe crabs waiting for the water to return. The sand is tinged with the green of horseshoe crab eggs. Thousands of eggs washing down with the outgoing tide are trapped by gravel on the beach. I'm inhaling no-see-ums (biting midges). Pretty, calico-patterned ruddy turnstones are digging for horseshoe crab eggs. An overpowering smell of decaying fish hangs in the air. Dead, dried eggs cover the beach: black grit, Weber calls them.

Windrows of fresher eggs line the sand at our next stop, Pickering Beach. The eggs feel rubbery. I can barely taste them, but they leave a sour, fishy residue. We turn over more stranded horseshoe crabs. A sign encourages visitors, "Just flip 'em," a project intended to cultivate compassion toward an animal whose ultimate survival, like that of many shorebirds, rests in our hands. Over the years, one woman taking morning beach walks has reported flipping thousands of horseshoe crabs. At a local bed and breakfast where I sometimes stay, the guests include physicians from California who come to Delaware beaches every year to flip crabs.

Not every beach we visit is sandy. We end the morning at Port Mahon. The beach here is rapidly retreating, moving over the road and into the marsh. Beach or no beach, horseshoe crabs are still spawning. Riprap and rock bolster the road, trapping the crabs, some wedged in so tightly they can't budge. It takes 10 or 15 minutes to pry the heavy rock away to release one tired horseshoe crab. Green eggs are piled in crevices between the rocks. The beaches are eggy today, but tomorrow the wind will change. "A few days of wind," Weber admits, "could wreck everything." Back at the Delaware National Estuarine Research Reserve on the St. Jones River, Weber and his crew will count eggs. Across the bay, New Jersey researchers are also counting eggs. Are there enough?

Knots flying in from Patagonia and Tierra del Fuego arrive emaciated. A 2,000-mile journey lies ahead. To make their flight to the Arctic with reserves to weather late-spring snows and undergo the physiological changes necessary to successfully breed, they must, in their short stay, increase their weight by 50 to 100 percent. They aren't all eating horseshoe crab eggs. South of crowded Delaware Bay, off the Virginia coast, lies a string of barrier islands, sandspits, and shoals backed by shallow lagoons and mudflats, the longest expanse of coastal wilderness along the U.S. eastern seaboard. Storms repeatedly pound the beaches—about 40 each year—cutting into the islands and shifting sand. Horseshoe crabs stay away from the rough waves and surf.

According to Barry Truitt from the Nature Conservancy, who has delved into the history of knots on the Virginia barrier islands, the written record of knots passing through there goes back at least 145 years, when ornithologists began collecting birds on Cobb's Island for museums. Truitt sent me one 1879 account, from ornithologist William Brewster of Cambridge, Massachusetts, who observed that knots were common on Cobb's Island in the spring, congregating on the beach and feeding on small black mussels. In his book *Seashore Chronicles,* Truitt offers an observation from actor, lawyer, novelist, and hunter Thomas Dixon Jr., who had a cabin on the island and wrote in 1895 that he saw flocks of "ten thousand" red-breasted snipes, "chattering and feeding" on the beach, where they fed "almost exclusively" on mussels. Dixon, Truitt points out, regularly took the overnight train from New York to Cape Charles, Virginia, and then the boat to Cobb's Island. The famous hotel, excellent seafood, and great birding and hunting on Cobb's Island attracted sportsmen and ornithologists from all along the eastern seaboard.

Today, some 13,000 knots stop in Virginia in the spring. Truitt, resighting knots on Hog Island one spring, found an international gathering: "two flags from Delaware Bay, one from Hog Island, two from Grande Île in Québec, one from Argentina, and one from Bahía Lomas. The birds all looked thin from their long distance flights." Almost half the knots in Virginia, 43 percent, come from Chile and Argentina, suggesting that these long-distance migrants refuel on the Virginia barrier islands as well as in Delaware Bay.

Knots spend 11 or 12 days in Virginia. They still eat blue mussel spat growing in banks of peat exposed as beaches migrate over the marshes, toward the mainland. They eat tiny clams as well. Since 1995, the spring population of knots on Virginia's barrier beaches has been stable, but in the coming years, as the water continues to warm, their blue mussel prey may disappear. In only 50 years, the southern end of its range has shifted more than 200 miles north from Cape Hatteras to Lewes, Delaware, at the mouth of Delaware Bay. As mussels continue moving poleward, right now at a rate of four miles per year, eventually Virginia's peat banks will be beyond the reach of larvae drifting south. Truitt thinks that an abundance of mussels in cooler, deeper water offshore could have supplied the good mussel set he saw in the springs of 2013 and 2014.

In Delaware Bay mussel spat and tiny clams are not abundant, and larger clams and mussels are too big to swallow, their shells too hard to digest. Knots filling up on the smaller shellfish need time to crush the inedible shells in their gizzards. Those heading north from Europe make a three-week stopover in Iceland and eat periwinkles, gaining weight at only half the rate of knots eating crab eggs in Delaware Bay. A periwinkle diet suffices; the Arctic is only a short hop away. In Virginia, Truitt has observed knots "so full they can't eat any more." At other times they forage day and night, and for as long as 18 hours a day, to gain the necessary weight.

Delaware Bay provides higher-quality, energy-rich food, attracting hundreds of thousands of sandpipers. Sandpipers love horseshoe crab eggs. Ground with a little sand, they fill the guts of dunlin, ruddy turnstone, knot, sanderling, and semipalmated sandpipers feeding on bay beaches. Shorebirds in Delaware Bay feed primarily, perhaps exclusively, on horseshoe crab eggs, congregating near the mouths of creeks where eggs accumulate. Scientists don't catch and kill birds the way they did in the nineteenth century to determine what they're been eating; they use more indirect means. Each method—flushing the guts of shorebirds, looking for nitrogen signatures in their blood or fatty acids in their tissues—speaks of a diet of horseshoe crab eggs.

Ounce for ounce, lipid-rich horseshoe crab eggs contain six times more energy than low-fat, high-protein mussels. Birds easily digest the eggs, turning as much as 70 percent directly into fat. Piersma found that "shorebirds as a group have unrivalled capacities to process food and refuel fast," but knots eating horseshoe crab eggs store energy at rates that are among the highest recorded in the animal kingdom. Each spring, a spawning horseshoe crab lays some 80,000 eggs. To gain the necessary weight, a single knot must consume 400,000 eggs; 40,000 knots need 16 billion eggs. Put another way, scientists calculate that ruddy turnstones, red knots, sanderlings, semipalmated sandpipers, and dunlin now migrating through the bay must eat 330 tons of eggs.

Horseshoe crabs are affected little, if at all, by shorebirds eating so many eggs. Most horseshoe crab eggs die: out of every 100,000 eggs, one crab larva will live a full year. The egg clusters rest deep in the sand, beyond the probing beaks of most shorebirds. They find eggs at the surface, those churned over by waves or exhumed by female crabs plowing through each other's nests. They would dry out if birds and tiny fish didn't eat them. Back Beach in Mispillion Harbor is the mother of all spawning beaches. Today, egg densities on other Delaware Bay beaches are measured in thousands or tens of thousands per square yard: in Mispillion Harbor and occasionally on Moore's Beach in New Jersey, they soar into hundreds of thousands. Mispillion Harbor is the one place in Delaware Bay always awash in eggs. It's not enough. To support Delaware Bay shorebirds, more Delaware Bay beaches need more eggs. In 1991, New Jersey beaches were thick with eggs—100,000 eggs per square yard on Reeds Beach, with stretches of sand packed with 300,000 eggs per square yard. Moore's Beach had even more eggs—stretches of sand with 500,000 eggs per square yard. By 2005–7, only a vestige of this abundance remained: egg densities on New Jersey beaches averaged only 4,000 eggs per square yard.

Delaware Bay, the most important refueling stop for red knots in eastern North America, is a critical rung on the red knot ladder between Tierra del Fuego and the Arctic. Breeding conditions in the Arctic, as I would find out, can be less than optimal. The birds must be fattened up and ready to take advantage of summers when conditions are favorable. Without enough horseshoe crab eggs to eat, shorebirds left the bay skinny. In 1998, almost 90 percent of Delaware Bay knots departing for the Arctic carried enough fat to reach the breeding grounds and, if necessary, weather an inhospitably late spring and still produce viable eggs. Four years later, the percentage had dropped by two-thirds. The number of adult knots surviving from year to year, once as high as 90 percent, plummeted to 56 percent, a rate far lower than in other knot populations. All the eggs, so to speak, were in the Mispillion Harbor basket, and the tiny beach couldn't supply enough eggs.

Knots weren't the only shorebirds in Delaware Bay getting a light meal. Semipalmated sandpipers, birds distinguished by tiny webbing in their feet—usually impossible for me to see—are gaining less weight than they once did, and gaining it more slowly. In Delaware Bay, their numbers have dropped a disconcerting 75 percent. Ruddy turnstones, birds I'd come to love watching as they staked out a claim over a cluster of horseshoe crab eggs, may not be finding enough eggs. Along the East Coast of North America, their numbers are dropping steadily and substantially, declining by more than half since 1974.

Shorebirds in Delaware Bay weren't getting enough food. Horseshoe crabs, once again, had been besieged. Niles noticed it when he began seeing tractor trailers parked at New Jersey's spawning beaches, loaded with horse-shoe crabs. He wasn't the only one. I heard stories of big trucks parked on the roadside filled with bleeding horseshoe crabs. It wasn't pretty. Fishermen took thousands of horseshoe crabs from the bay each year, then tens of thousands, and then hundreds of thousands. By the mid-1990s, they were killing almost 2 million horseshoe crabs every year, taking these "trash fish" to use as bait.

No wonder I never saw the profusion of eggs that once characterized the bay. I never saw eggs so thick they could be carted away in wagons, never saw Wilson's bushels of eggs. By the 1980s, after the fertilizer industry collapsed and the crabs had begun rebounding, Myers reported "whole stretches of beach . . . covered with rafts of wave-washed eggs," and in "a few places where currents converge, the windrows pile up to a thickness of two feet." I never saw this either. I walked almost every bay beach where horseshoe crabs spawn. At most I found a few teaspoons—a small pile of eggs wedged in the rock and riprap at Port Mahon.

The lessons of the horseshoe crab fertilizer industry long forgotten, history was repeating itself as fishermen depleted the bay of horseshoe crabs and shorebirds went hungry. The Atlantic States Marine Fisheries Commission (ASMFC), responsible for managing the horseshoe crab fishery, hadn't been managing: during the 1990s, the agency hadn't restricted the number of horseshoe crabs fishermen could take. Biologists and conservation organizations, worried about the steep decline in shorebirds, stepped in to stop the decimation. At their urging and insistence, the Delaware Bay states, the U.S. Fish and Wildlife Service, and the ASMFC began taking steps to stanch the decline of horseshoe crabs, recognizing in 1998 that horseshoe crabs were disappearing. Fishermen, fisheries managers, and wildlife biologists didn't always agree on the remedy: tensions ran high, negotiations were contentious. The debate went on for years. By 2006, regulators had established a moratorium on the New Jersey horseshoe crab bait fishery; instituted catch restrictions in other Delaware Bay states, including a prohibition on taking female crabs at any time; closed the male horseshoe crab fishery between January and early June; introduced alternative bait; and established a horseshoe crab sanctuary at the mouth of the bay.

These actions were designed to reduce the take of horseshoe crabs for bait by 70 percent. The argument about how much to limit the horseshoe crab take continues, often heatedly. In 2014, the New Jersey legislature considered and rejected a proposal to end the prohibition on the bait fishery. I heard people say that Niles should stick to science and stop advocating. If his colleagues and he don't advocate, who will? Shorebirds and horseshoe crabs have no voice, and, as the aftermath of Hurricane Sandy would reveal, horseshoe crabs face many threats. Shorebirds aren't the only ones whose health and well-being depend on the health and well-being of horseshoe crabs. Ours do as well. Without them, there'd be fewer of us.

Six

BLUE BLOODS

I n the warren of research labs, hospital buildings, and parking garages that make up Massachusetts General Hospital's (MGH) Boston campus, a building where radioactive pharmaceuticals are manufactured barely stands out, even though it hosts a multimillion-dollar research and clinical program in nuclear medicine. The building, characterized by old brick and dirty windows, and seemingly without an easily identifiable name or number, backs onto the sidewalk. The one street-side door opens only to staff swiping IDs. No one I ask for directions knows of it. I finally find an unlocked entrance at the end of an alley, adjacent to a pretty courtyard and patio. Underground, below the tables and chairs, a particle accelerator is encased in a six-foot-thick vault of concrete. About the size of a small garden shed, it weighs as much as 22 elephants.

One of two cyclotrons at the hospital, it's been bombarding atoms since 3:30 a.m. Inside the accelerator, high-energy beams of subatomic particles spiral through an electromagnetic field at 25,000 miles per second. It's fusing atoms to make tracers that, when injected into a patient undergoing a PET scan, allow researchers to track the progression of cancer, heart disease, and degenerative diseases such as Alzheimer's and Parkinson's. These new tracers, like the uranium and plutonium isotopes in atomic bombs and nuclear power plants, are radioactive. Unlike them, the lives of these new tracers are fleeting. Plutonium 239, used to manufacture nuclear weapons, endures: its half-life is 24,100 years. Fluorine 18 (F-18), a radioisotope made at MGH, disintegrates quickly, its half-life a mere 109.77 minutes. Technicians, chemists, and nuclear pharmacists have only a few hours to create the atomic particles, synthesize the tracer, test it for purity, inject it into the patient, and run the scan. If they are delayed, the radioisotope breaks down, rendering the tag useless. If bacteria contaminate the tracer, the patient could succumb to a raging infection. I'd come to see the unique role horseshoe crabs play in detecting potentially lethal bacteria.

Microbes, the first life to evolve in the ocean 3.8 billion years ago, are everywhere: more than a million bacteria fit in a quarter teaspoon of water. Bacteria are also everywhere—in soil, in the air, and in water, even drinking water. The human body contains an estimated 100 trillion cells. Bacteria outnumber human cells 10 to 1: they cover our skin; they live in our noses, throats, sinuses (where some 1,200 species reside); and in our intestines (where 99 percent of the genes are bacterial, not human). Take away the bacteria, and little would remain. Bacteria are part of who we are. They are critical to our health: helping to digest food, fight infection, and stave off asthma. Many are harmless, but some release powerful poisons, causing pneumonia and meningitis, food poisoning, or whooping cough.

Before drugs and medical devices enter the human bloodstream or spinal fluid, harmful bacteria they may contain need to be identified and disarmed. The test for bacterial contamination in injections and medical equipment is ubiquitous, applied not only to radioactive tracers and not only at MGH, but also at hospitals and pharmaceutical companies around the world. It is used on vaccines, syringes, IV drugs, IV lines, stents, hip and knee replacements and other medical implants; on drugs injected into human muscle, blood, bone, or subcutaneously. In other words, all of us, and our pets, benefit from this test. Its accuracy is imperative. Most of us, our children, or our parents, and perhaps our dogs and cats, have had blood drawn or received IV antibiotics or rehydration. If the lines and needles weren't sterile and free of toxic bacteria, many more of us would experience antibiotic resistant infections, some of them fatal.

In the early days of intravenous therapy, many people did die. Intravenous therapy, now routine, was once frightening and risky. When cholera swept across Russia and Europe in the early nineteenth century, a few physicians desperate to find a treatment injected water, saltwater, and/ or laudanum into their patients' veins. The results were poor. Patients died within minutes or hours. In 1832, Dr. Thomas Latta administered this infusion to 15 patients. Ten died; this was considered a success. In 1847, another physician, J. Mackintosh, before injecting IV fluids into 156 patients, "carefully strained" his water solution through leather, a precaution that probably made matters worse: 84 percent of his patients died, a much higher rate than those spared this innovation.

With death rates like these, it was difficult to consider injections and IV rehydration any different than other dangerous and ineffective treatments: castor oil and milk of magnesia cathartics, calomel purgatives, and bloodletting (as much as a quart every two hours) from already severely dehydrated patients. The British medical journal the *Lancet* called these irrational treatments "full of sound and fury, signifying nothing." To other critics, they constituted "benevolent homicide."

Though hard to recognize at the time, intravenous therapy would become the one therapy that could treat cholera and save lives, reducing the 70 percent mortality rate among hospitalized patients, but it would be a long time before physicians determined the right balance of salts for the injections, and even longer before they could deliver them safely. Intravenous therapy grew in popularity, but as it did, so did the risk of associated and sometimes fatal fever. Salvarsan, the arsenic-based drug synthesized in 1910 to treat syphilis, rapidly became the world's most widely prescribed drug. Problematically, salvarsan injections induced "salvarsan fever." Surgical IV drugs induced anaesthesia fever and surgical fever. Each new drug appeared to induce its own unique fever: protein fever, seawater fever, water fever, sugar fever, tissue fever.

A young biochemist, Florence B. Seibert, found this puzzling. Ten years earlier, two men had posited the novel idea that contaminated water was the culprit. Their idea was ignored, their paper neglected and forgotten until Seibert began her investigation. In 1923, she published the surprising findings of her Ph.D. thesis, proving that although each fever had a different name, they shared a common source—bacterially contaminated water in the injection fluids. Although physicians presumed that distilled water must be safe, she proved otherwise, showing it could easily be contaminated by bacteria that remained stable even when subjected to heat. These bacteria could induce not only fever but also headache, chills, nausea, and death. Isolating them was critical.

Seibert's pioneering work dramatically improved the safety of IV therapy and is probably among the most useful Ph.D. dissertations ever written. She pinpointed the cause of "injection fever" and developed the means to prevent it, designing a baffle to trap bacteria caught in water as it was distilled. Her work, which, according to one colleague, had "seemed of moderate interest at the time ... later proved of vital importance in the broad fields of intravenous medication and blood transfusion." Seibert further applied her findings, refining and perfecting a way hospitals and the newly emerging pharmaceutical industry could test IV drugs for purity. Cradling rabbits during the day and feeding them oats and cabbage at night, Seibert found that their temperatures fluctuated little. If she injected contaminated water into veins near their ears, they spiked fevers. The rabbit test became the FDA gold standard to evaluate the safety of injected drugs. Her research, which she conducted at a time when universities relegated women to the lowest-ranking and most poorly compensated positions, enabled the pharmaceutical industry to supply safe IV equipment and fluids to hospitals and to wounded soldiers fighting overseas in World War II.

These accomplishments would seemingly have more than satisfied the most ambitious researcher, but the indefatigable Seibert continued on, publishing 119 scientific papers and solving another intractable problem plaguing scientists. In 1890, Robert Koch had identified the germ that caused tuberculosis but he failed to fully isolate it. His inability to rid the bacillus of impurities rendered the diagnostic skin test wildly unreliable. After 10 years of effort, Seibert succeeded where he had failed, straining out the impurities with an explosive—a sticky piece of guncotton—placed over a porous clay filter. The World Health Organization formally adopted her tuberculin test; it is still the standard skin test used today. The rabbit test would endure for 50 years. When scientists developed a better one, its source would be another animal: one that doesn't eat hay and oats, and certainly doesn't cuddle on anyone's lap.

Important scientific discoveries have been hastened, possibly enabled, by the proximity of the Marine Biological Laboratory in Woods Hole,

Massachusetts, to beaches and marshes where horseshoe crabs spawn. Summer in Woods Hole begins with the simultaneous arrival of inquisitive scientists and spawning horseshoe crabs, and the crab has become a popular and valued subject of investigation.

Biologist H. Keffer Hartline loved the huge light receptors he found in horseshoe crab eyes. Horseshoe crabs, *Limulus polyphemus,* were named after Homer's giant Cyclops Polyphemus, for the two eyes located in the center of their shell, like the eye of the Cyclops. These eyes are tuned to the light of the moon and the stars. Polyphemus had one eye: *Limulus* has 10. On the sides of its shell lie two more eyes, each with 1,000 photoreceptors, the largest in the animal kingdom, 100 times larger than our retinas' rods and cones. Another light sensor is on the tail, and the rest lie beneath the shell. Hartline examined how horseshoe crab eyes send electrical impulses to the brain in response to light. He laid the foundation for understanding human vision. His work on horseshoe crab eyes revealed how animal eyes respond to different intensities of light, and how we perceive contrasts. For this research he was awarded a Nobel Prize.

From the time dusk falls until the sun rises, horseshoe crabs' sensitivity to light increases by a factor of 1 million. They see at night as if it were day. Photoreceptors along the crab's tail send signals to the brain, synchronizing this circadian clock to the actual rhythms of light and dark. Other eyes detect light below the crab or help baby crabs find their way as the rest of their eyes mature. Despite the crabs' having so many eyes, Hartline, unable to ascertain what they actually saw, joked that he spent years "studying vision in a blind animal." His student Robert Barlow, who'd discovered the circadian rhythms in horseshoe crabs, donned scuba gear to try to learn what horseshoe crabs are looking for in the water, but "after many cold and lonely nights at the bottom of [Buzzard's] Bay," he learned only that the animals avoided overhead shadows the moonlight cast from his underwater clipboard. Later, lining the high tide line with cement castings of female crabs alongside similarly sized cubes and half circles, and waiting for the night's high tide, Barlow and his colleagues came to understand that horseshoe crabs have eyes primarily for each other.

Hartline and Barlow studied horseshoe crab vision. For other researchers, the ungainly crab would yield a precious pharmaceutical. Gram-negative bacteria are of great concern in the production of drugs and medical devices. Fragments of their cell walls can contain dangerous toxins that survive sterilization, remaining potent even when the bacteria themselves are killed. Florence Seibert couldn't identify the endotoxin contaminating her distilled water—she called it a "little blue devil"—but her feverish rabbits signaled its presence. The blood of the horseshoe crab would prove a better indicator.

In 1950 and 1951, Frederick Bang, studying immunity at Woods Hole, discovered that the blue blood of horseshoe crabs clots in the presence of Gram-negative bacteria. The clotting agent is so powerful that before a horseshoe crab would succumb to a lethal infection, all its blood might clot first. Much time can elapse between a scientific discovery and its application. It would be another twelve years before Bang and another colleague from Johns Hopkins, hematologist Jack Levin, together isolated and stabilized the crab's highly sensitive endotoxin detector, called LAL: L for *Limulus;* A for amebocyte, the horseshoe crab's blood cell; and L for lysate, the toxin detector that remained after scientists separated the horseshoe crab's blood cells from the plasma and burst them apart.

A few years later, James F. Cooper from the U.S. Public Health Service enrolled at Johns Hopkins to do graduate work. His interest, the newly emerging field of nuclear medicine, would mark the beginning of the end of the long-standing rabbit test for fever-inducing bacteria. Rabbit tests were expensive. Large numbers of rabbits required housing, care, and trained personnel to handle them. Using confined animals for drug testing raised ethical issues. The rabbit test was far too cumbersome and time-consuming for short-lived radiopharmaceuticals coming into use for medical imaging. As Allen Burgenson, manager of regulatory affairs for Lonza, a company that makes LAL, explained to me, "Rabbits are skittish. They will sometimes spike a fever if a stranger walks into the room. They have to keep calm. They need to be only with people they know. If you have a $1 million drug, you can't even let people in the room where it's being tested."

Furthermore, the rabbit test didn't always work. In patients suffering head trauma, fluid can leak from the protective cushion surrounding the brain. Doctors injected radioactive tracers into the patient's spinal fluid to diagnose this condition. The new tool carried serious risk. Endotoxin is hundreds of times more toxic when injected via spinal taps than intravenously, and rabbits failed to detect these smaller quantities. Cooper immediately saw LAL's potential. In one 15-month period, he found that large numbers of patients, as many as 27 percent, developed fevers from the tracers, and as many as 14 percent developed meningitis, a serious adverse effect that can result in stroke or death. The more sensitive LAL picked up the contamination.

Cooper and his colleagues went on to explore the sensitivity of LAL, extracting toxins from *E.coli* and *Klebsiella* bacteria normally found in the human intestine and subjecting rabbits and crabs to different concentrations. *Klebsiella,* an increasingly antibiotic-resistant source of hospital-associated infection, can cause meningitis, bloodstream infections, pneumonia, and surgical infections. The researchers found that horseshoe crab blood was 10 times more sensitive to endotoxin than the rabbit test, even signaling the presence of dangerous endotoxin when rabbits showed no reaction. Still, LAL was new and relatively untested, and pharmaceutical companies were reluctant to switch from a proven, tried-and-true test. By 1985, after thousands of head-to-head LAL and rabbit tests, the FDA had approved LAL as a substitute test for fever-inducing bacteria, revolutionizing how injectable drugs are evaluated for safety.

Baxter, a pharmaceutical company in Chicago that manufactured intravenous drugs, began producing LAL, but horseshoe crabs didn't live in Lake Michigan's freshwater. In Beaufort, South Carolina, Robert Gault had plenty of horseshoe crabs. In addition to shrimping, softshell crabbing, and clamming, he'd take horseshoe crabs to New York in a tractor trailer and sell them for bait to conch and eel fishermen. One day, he can't remember exactly when, he was pulling his young son Jerry through Lucy Point Creek in a homemade surfboat when a stranger arrived. Suspecting an insurance salesman—the man was wearing a suit and tie—Robert Gault initially refused to come out of the water, but the unwanted guest, a representative from Baxter, had come to make an offer he couldn't refuse. "I was getting peanuts for bait," he remembers. "That man offered me a golden spoon. The company paid me to collect crab and fly them to Chicago, wrapped in newspaper and packed in cardboard boxes." Later, when Baxter opened a facility in South Carolina, the company housed its lab at Gault Seafood, with Jerry's mother, Blanche, as head technician, bleeding crab in the office, between 50 and 100 a day.

The researchers also went into business. Stanley W. Watson, a microbiologist at the Woods Hole Oceanographic Institution, reincarnated a real estate company he'd been running out of his garage, Associates of Cape Cod, to produce LAL. Cooper, who'd been bleeding crabs and making LAL for the federal government on Wallops Island, near Chincoteague, Virginia, set up shop as well. Hearing that horseshoe crab were big and plentiful in South Carolina, he moved his family to Charleston where, speaking with fishermen, he learned that tens of thousands of horseshoe crab spawned on the sandspits of Bulls Bay, just north of the city. These crabs were the largest he'd ever seen, and the water was clean. He learned that to the south, near Beaufort, there were even more crabs, spawning in the sheltered tidal creeks of Port Royal Sound. He planned to put his plant on Sullivan's Island, at the mouth of Charleston Harbor, but on September 21, 1989, Hurricane Hugo flooded the island, dislodging the connecting bridge to the mainland. Cooper chose a less picturesque location on higher ground in a nondescript suburb near the airport. Water quality in the bay was so bad in the hurricane's aftermath that in the beginning he didn't get many horseshoe crabs.

The company that Cooper founded is now part of Charles River Laboratories. Barbara Edwards, training manager at the company's Endotoxin and Microbial Division in Charleston, and her college roommate were Cooper's first employees. They cleaned and validated equipment, placed purchase orders, bled crab, whatever needed doing. Cooper bought used equipment from a pharmaceutical plant closing in Georgia. He cultivated investors but was undercapitalized and barely made it. As cases of patient infection from drugs that passed the rabbit test continued to occur, and FDA requirements tightened, pharmaceutical companies slowly adopted the new LAL test. Today, four large multinational biomedical companies make LAL from the blood of horseshoe crabs: Charles River Labs, operating in Charleston and also bleeding crab in Delaware Bay at the old King Crab Landing in Cape May Courthouse; the Swiss pharmaceutical company Lonza, working in Chesapeake Bay; the Japanese company Wako, bleeding crabs in Cape Charles, Virginia; and Associates of Cape Cod, now part of the Japanese firm Seikagaku, bleeding crab from Cape Cod, Massachusetts, and Rhode Island.

John Dubczak is general manager at Charles River's Endotoxin and Microbial Division. His eyes are as blue as the blood of a horseshoe crab.

While we talked, I gulped water to quench my thirst in the heat. He tested it with the company's new LAL cartridge. The test read positive, but the level was far too low to be of concern. Today's advances in medicine and surgery, which expose the body's cardiovascular and lymphatic systems, blood, and spinal fluids to drugs and medical devices, require endotoxin tests to be highly sensitive. The body, wrote Lewis Thomas in *The Lives of a Cell,* treats Gram-negative bacteria as "the very worst of bad news." Sensing endotoxin, "we are likely to turn on every defense at our disposal; we will bomb, defoliate, blockade, seal off, and destroy all the tissues in the area." The result, he says, "is a shambles," the consequences including fever and inflammation, low blood pressure, respiratory distress and suffocation, shock, and death. To avoid the body's defenses running amuck fighting endotoxin, standards governing drugs, vaccines, intravenous solutions, and medical devices limit exposure. The LAL assay determines whether those standards are being met.

During the high tides of May, water pours into the creeks and estuaries of the South Carolina Lowcountry. Low means low; the coast, softened into wet, marshy islands by rivers and tides, belongs more to sea than land. The sea is everywhere: in the dank smell of the marsh; on roads that barely edge the upper reach of tides; in fertile creeks teeming with shrimp and crab; in light-filled vistas that begin and end with water. On small islands barely rising above the sea, on this constantly changing, ephemeral landscape, the claim of land is tenuous and temporary.

At the edge of Lucy Point Creek stands the home of Gault Seafood. There's the family house, a few low-slung buildings, some trucks, and a sign in front: "Parking for fishermen only. All others will be thrown back." I hesitate, but Jerry Gault, now grown up, is expecting me. Inside, a woman buffs claws from stone crab. They glow a beautiful brownish pink. A delicacy reputed to be far tastier than lobster, stone crab live in mud holes at the edges of tidal creeks. In the 1880s, fishermen took them "by the hand, thrust down several inches, sometimes fifteen to twenty, to reach the inhabitant at the bottom, at the risk of a severe bite from one or both of its claws." Nowadays, Gault uses a trap, but the ferocity of this pugnacious animal hasn't abated. "There is one local fisherman who lost his finger to stone crab. . . . The trick to not getting pinched is how you pick them up and how fast you do it. It's real important to be faster than the crab. I had to let one of my helpers go because he kept on getting pinched. I could not stand to see his pain. I was worried he was going to lose a finger."

In another room, cool, freshly filtered creek water flows into tanks filled with blue crab practically busting out of their shells. After the molt, and when they are barely firm, Gault will ship to his buyers. In the office, crab steam in a large pot. I'm offered a plate; I've been eating crab all week, listening to stories of she-crab soup with cream and sherry on the side but this—all succulent meat, unadorned, sprinkled only with the slightest pinch of pepper—is the sweetest and the freshest. In the water, stone and blue crab fight. Gault hauls his traps to find them at a standoff, locked in each others' claws, a death embrace. Tonight he seeks a crab that neither puts up the stout resistance of stone crab nor tastes as heavenly as blue crab.

Still, when I step into the boat, Gault hands me a pair of gloves, and I notice he is wearing rubber boots and that he's built a makeshift knee-high plywood wall around the wheel. The tide's so high that Gault made two runs for horseshoe crab yesterday, getting out a second time before the water ebbed. Tonight, high tide is later, around 10:30, so he'll make just one run. He predicts the horseshoe crabs will start coming in at 7:50 this evening, and he is expecting "a massive orgy."

As the sun sets, we head out into Wallace Creek, Chowan Creek, and the Broad River, into the vast labyrinth of named and nameless Lowcountry creeks and mudflats that appear and disappear daily with the tide. The water widens, the land softens into a darkening silhouette, and eagles appear in the trees. The crew changes into wetsuits. We pass a large beach at a creek mouth. It's empty. "You'd think this beach'd be great for crabs, but we rarely pick 'em up here," Gault says. "There's no reason why they're in one place or another."

What he does know after 17 years in the business is that crab seem to be, if not more numerous, a little larger. Gault gives three reasons, the first that "shrimpers now use turtle excluders." This gear, designed to spare sea turtles from shrimp nets, also frees horseshoe crabs. "And there's no longer a bait fishery for horseshoe crab, which wasn't all that big anyway. Shrimpers would bait the bottom with dead horseshoe crab; now they use a mud ball mixed with dried menhaden. Lastly, you can't drag the river here anymore. With the trawling ban, prime bottom habitat isn't getting dug up." This is all good news for Jerry: 20 percent of his income comes from selling horse-shoe crab.

We pass a small, low point covered with birds. "You can tell there's crab there." Gault gestures toward the shorebirds clustered on the beach. The point, though, is about to disappear under the rising water. In another year it, like so many shoals and spits along this evanescent shore, may disintegrate entirely. "These places change so much. From one year to another, they build up and erode away." Maps quickly become outdated. Gault has two plotters and a depth sounder to guide him through the maze of creeks and rivers of this inconstant seascape. In the fading light, we approach an oyster bank somewhere along the porous edge of Parris Island. Gault lets me off, telling me to go explore while he scopes out crab. His boat disappears into the darkness; I've left my cell phone (which probably wouldn't work here anyway), wallet (also useless), and flashlight behind. I did look at one nautical map beforehand, but I don't have the vaguest idea where we are. It's a beautiful night. The trees are black; the oysters gleam. This is what Chesapeake Bay must have looked like 100 years ago, before we leveled the oyster banks and took all the oysters, and when the great reefs were hazardous to navigation.

The bank is full of crab. They've burrowed into the oyster shell everywhere, a bit of shell exposed here and there, and piles of bulldozed oyster shell marking horseshoe crabs buried below. Horseshoe crabs in the tidal creek where I live fit in the palm of my hand. These are more than a foot wide. Gault returns. Reaching over the bow to help me in, his white boots shining in the dark, he gestures into the blackness. "That point," he says. "We're going to that point." I can't see anything. He's excited. "I don't know whether it's the spawning or what, but there's something intoxicating, going out, picking up crabs." It is intoxicating—the warm heavy air, the inky darkness, the smell of salt and creek, the reams of oysters, and the moon, still to rise. Once a year, the full and perigee moon coincide. This year, the time when the oval orbit of the moon comes closest to the Earth is now. This month, the full and perigee moon are less than a minute apart, creating a "supermoon" that will shine 14 percent brighter and appear 30 percent bigger than other full moons. It hangs just below the horizon.

We reach the oyster bank. Gault peers over the gunwales. "They're coming now, they're coming now." His instincts are unerring; it's 7:52. The crew members don gloves and headlamps, unload the flat-bottomed john-boat, and wade in. The wind pushes the tide. Gault is a little worried. This tide, already nine feet, already rising high on the bank, may flood the spawning grounds. Right now, the water brims with horseshoe crabs heading in for the shell reef. It takes only 10 minutes for the boys to fill the johnboat. They empty horseshoe crabs into the big boat and refill the john-boat two more times. There's no dearth of horseshoe crabs. In less than two hours, we're standing up to our shins in horseshoe crab, ready to head back. About 4,500 pounds (about 900 horseshoe crabs) wriggle on the deck. A 250-horsepower engine on a 25-foot boat seemed excessive earlier, but now the boat won't plane. I listen to a discussion about whether we've got 40 crabs too many or too much weight from writers. Gault tells me about a Carolina skiff that sank under the weight of a generous haul. We turn into a creek that leads toward the landing, and the men of the crew pull out their cell phones and check Facebook. Huge rafts of ebbing marsh grass catch the motor. The moon rises, blood orange behind live oak.

Biomedical companies want the blood of Gault's crabs. During the weeks of the May high tides, technicians working for Charles River Laboratories extract blood from horseshoe crabs. When I arrive in Charleston at 8:30 a.m., the heat and humidity outside already verge on unbearable. At a dock in the back of the building, one fisherman unloads a trailer. He had taken horseshoe crabs from Cape Romain, north of Charleston, the night before, and brought them down early this morning. To minimize damage en route, when crabs can overheat or dry out, the company pays by the crab, inspecting each as it is unloaded, rejecting the obviously injured or lethargic. Three families of fishermen, including Gault and his father, supply horse-shoe crabs to Charles River Labs, and have been doing so for years.

Production is ramped up during the month of spawning. On busy days, work can begin at 6:00 in the morning and finish at 7:00 or 8:00 at night. It's busy this morning. Once in the building, the horseshoe crabs, scraped of barnacles, head for the bathing station, a long, deep, gray metal sink nicknamed the "spa." Here, they are measured, small ones rejected, the rest hosed off with cold water, dipped in disinfectant, and brought to a "staging area" where, belted to racks mounted on wheeled stainless steel tables, they await bleeding. The wait isn't long. A rack of males, finished, is wheeled out and a rack of females quickly wheeled in. Technicians wearing face masks, lab coats, and gloves, their hair netted, spray the crabs with alcohol. They insert 14-gauge needles into the membrane surrounding the crab's heart, and blood flows.

Human blood, containing iron, reddens in the presence of oxygen. Horseshoe crab blood, containing copper, turns milky turquoise as it flows into glass bottles. After 8 or 10 minutes, when the horseshoe crabs have been bled, the rack is whisked back out onto the main floor and the animals returned to the trailer, now under a tent shielding them from the hot sun. The fisherman waits. When all his horseshoe crabs are bled, he will return them to the sea.

It takes the blood of two, maybe three crabs to fill each bottle that, along with its foil cap, has been baked in a high-heat oven for three hours to kill any bacteria and destroy endotoxin. Still, every once in awhile, the blood coagulates, signaling contamination, and workers throw it away. It's this telltale response to toxin—clotting—that makes horseshoe crabs and their blood so valuable. If our immune system can't fight an infection, we have an arsenal of vaccines and antibiotics. Out in the ocean, the horseshoe crab rises to its own defense, its blood coagulating around invading toxic bacteria, immediately sealing it off to prevent systemic infection. Horseshoe crab blood cells immobilize, inactivate, and eventually destroy pathogens. A protein known as Factor C triggers the cascade of reactions that clots horseshoe crab blood in the presence of endotoxin. It is an immune system of exquisite beauty that, according to Cooper, may explain why horseshoe crabs endure. "Our most important predators are bacteria," he says. "We perish as the body works, sometimes overworks, to get rid of bacterial toxins, while crabs can deal with them."

Throughout the day, technicians turn blue blood into liquid gold. They centrifuge it to separate blood cells from plasma and then add sterile water. As the freshwater permeates the salty blood cells, the cells grow spongy and burst from osmotic pressure. The sloughed cell walls sink. Above them floats LAL, the toxin detector. It's further processed and tested for contamination, and then the faintly yellow-tinted liquid is freeze-dried and sealed.

The day I visit Mass General, the exquisite sensitivity of LAL is being put to use. Daniel Yokell is the manager of PET (positron emission tomography) production chemistry at the hospital's PET Center radiopharmaceutical manufacturing facility. Every day, he supervises the manufacture of between three and six batches of radiopharmaceuticals. The day I am there, Yokell, a nuclear pharmacist who's always had an interest in chemistry, and his staff—two technicians, one nuclear pharmacist, and two chemists—have programmed the cyclotron to bombard oxygen-enriched water, fusing oxygen with protons and neutrons to make F-18, a radioactive isotope of fluorine. They are also creating C-11, a radioactive isotope of carbon. Since its half-life is only 20 minutes, they must make more than they need for a scan.

Technicians monitor production from a bank of computers in a windowless room. Bombarding the atoms takes about 30 minutes for C-11 and up to two hours for F-18, depending on how much isotope they need. The radioactive material is piped under the floor into a lead hot cell where it is synthesized into tracers for PET scans. Technicians gown into garments designed for the ultra-clean environment required for the production of these drugs. Strips of blue sticky tape lining the floor grab dirt and grit from their shoes as they approach the room housing the cell. A lead glass viewing window shields them from radioactivity.

When my daughter was hospitalized with appendicitis, the blood of horseshoe crabs ensured that her IV lines and medicines were free of high levels of endotoxin. Over the years, each of my family members has been vaccinated: for whooping cough, tetanus, hepatitis, measles, and flu. Horseshoe crab blood ensured that these vaccines and IV medications we've received in the hospital were safe from endotoxin. In 2013, doctors inserted a stent to reopen a blocked artery in former president George W. Bush's heart. He returned home the following day with no complications or infection, partly because horseshoe crab blood ensures that medical devices implanted in the human body are free of endotoxin contamination.

The LAL endotoxin test is ubiquitous. The horseshoe crab, unseen and unknown to most of us, makes that test possible. In the winter of 2013–14 alone, 134.5 million doses of flu vaccine were distributed in the United States. Every year in the United States doctors implant half a million stents in patients with heart disease, manufacturers sell some 330 million intravenous catheters, 3 million people whose eyes are cloudy with cataracts are given new manufactured replacement lenses. Heart stents and intraocular lenses; hip replacements, pacemakers, and breast implants; injectable and intravenous antibiotics, chemotherapies, vaccines, and insulin; syringes and tubing used to administer them; and the water used in their manufacture and preparation: all these are made safe from endotoxin by the blue blood of horseshoe crabs.

We've come to take for granted the safety of our drug supply, but deviations from best-management practices and violations of endotoxin standards are reminders of horseshoe crabs' pivotal role safeguarding human health. Gentamicin is a powerful antibiotic used to treat severe bacterial infections. In 1998 and 1999, 155 patients experienced reactions to the drug, symptomatic of the presence of endotoxin. Some experienced chills, fever, shivering, high or low blood pressure, abnormally high heart rates, and low oxygen levels in their blood. One patient's lungs filled with fluid. While most recovered quickly, others were intubated, resuscitated, and admitted to intensive care units.

The cause: egregious violation of good management practices at a company in China producing bulk gentamicin for use by other manufacturers. The company failed to comply with endotoxin testing standards and did not perform endotoxin tests on water it used to make the drug or to rinse equipment. FDA tests revealed that some lots had unacceptably high levels of endotoxin, a problem compounded by doctors prescribing the antibiotic off-label, that is, giving it in one dose per day rather than three, thus increasing the patient's hourly exposure to endotoxin. The FDA banned the firm from importing into the United States, and the standards for endotoxin levels in gentamicin were tightened.

The use of radioactive tracers, like those made in Yokell's lab, is growing. A friend being treated for ovarian cancer at MGH couldn't know her chemotherapy's effectiveness until well into her treatment, when a CT scan would show whether her tumor had shrunk. That wait may shorten. "Radioactive tracers are fundamentally changing how oncology is managed," Yokell tells me as we walk through his lab. "Hypermetabolic tumors, taking up abnormal amounts of glucose, will trap fluorine-labeled glucose, enabling physicians to visualize a tumor's molecular changes and therefore determine weeks and months sooner whether a cancer is responding to chemotherapy." My friend went through multiple rounds of chemotherapy before she could learn her drug wasn't working. A metabolic analysis of her tumor, now most often applied to bone marrow cancers, might in the future direct other cancer patients early on to the most effective treatment. Horseshoe crab blood makes use of these new tests possible.

Ongoing research suggests other potential applications for radio-pharmaceuticals. The day I'm in Yokell's lab, a tracer will image receptors in the brain of a patient with Alzheimer's, a disease affecting 5.2 million Americans over the age of 65 (one in eight). Another 1.3 million suffer from Lewy Body Disease, another form of dementia whose origins, like that of Alzheimer's, are poorly understood. My father died of Lewy Body Disease. An avid skier and tennis player, he could barely shuffle across a room as the disease took hold. His lucidity waned. On what I reluctantly came to recognize as his good days, he'd sit contentedly in his favorite chair, enjoying the *New York Times,* holding it upside down. Or he'd gaze fondly across the living room, chatting with a beloved Irish setter who'd died years earlier. On bad days, confused and frustrated, he'd puzzle over his bank statements, asking me repeatedly to explain the difference between 18, 180, and 1,800. It was heartbreaking. Yokell thinks that new research using radioactive tracers may eventually help locate the origin and track the progression of these debilitating dementias. When my father was younger, and his mind and body were more agile, he and his wife rented a house in Falmouth, Massachusetts. He took my children, then toddlers, to a nearby beach packed with spawning horseshoe crabs. Watching the ungainly animals, neither he nor I realized that someday they might make it possible to understand the disease that would kill him.

I'd arrived at MGH at 11:30 a.m. It's now 1:00, and Yokell and I have one more stop. After the radioactive tracer is synthesized, sterilized, and purified, it goes to quality control where it is tested for, among many other things, pH, clarity, impurities, and residual solvents used in the manufacture of the drug. Some 30 pages of check-offs alert Yokell to the source of any problem. He pulls out the LAL cartridge and its endotoxin results. The cartridges are expensive, $40 each, but they are fast. Older tests took an hour. This test yields results in 15 minutes. The standard for this particular drug, according to Yokell, is 17 units of endotoxin per milliliter. The test registers less than 1. The patient is waiting. The drug is good to go.

Seven

COUNTING

On shore, when the tide is high and the moon is full or new, scores of scientists and volunteers all along the eastern seaboard, from Florida and Georgia up through Long Island and into Massachusetts and New Hampshire, are counting horseshoe crab coming in to spawn. Out in the water, scientists aboard old fishing boats are trawling the seafloor, counting fish, shellfish, jellyfish, and horseshoe crabs. From beaches and from airplanes, they count shorebirds. They also track the rate of erosion on bay beaches and calculate the number of truckloads of sand necessary to restore them. Everyone needs to keep tallying.

In 2004 the highest number of knots scientists counted in Delaware Bay was 13,000. In 2012, it rose to 25,000, a far cry from the 95,000 Dunne saw in the early 1980s, but a hopeful improvement. (Because knots can depart from the bay before the peak count, or arrive afterward, the total number migrating through is higher. Modelers have estimated these numbers: 17,000 in 2004, 45,000 in 2012. There are no comparable estimates for the 1980s. Some scientists put the number at 150,000.)

The peak counts held in 2013 and 2014. It appeared as if years of dogged persistence in tightening regulations restricting the take of horseshoe crabs had begun to stanch the downward spiral of horseshoe crabs and knots in the bay. The thin rise of sand in Mispillion Harbor continued to supply masses of eggs for migrating shorebirds, and egg concentrations on a few other beaches were beginning to rise, in part due to beach restoration after Hurricane Sandy. More birds—both semipalmated sandpipers and knots—were gaining the weight they needed for the long flight north. In 2013, 46 percent of knots left the bay plump; in 2014, 53 percent.

These encouraging numbers are a welcome relief. It's too soon, however, to know whether they will persist. Conditions in the bay during the 2012–14 shorebird migration were particularly favorable. There were no major storms, no high-speed westerly winds churning the waves and limiting spawning to protected creeks, no cold snaps delaying horseshoe crab spawning. One change in the weather can upset the delicate synchrony between birds and crabs. In 2008, a nor'easter blew in high winds and waves. Birds arrived, but horseshoe crabs took shelter out in the water. Egg densities on Delaware Bay beaches peaked only after knots had departed: fewer than 15 percent gained the weight they needed. In such a year, Larry Niles has seen as many as 4,000 birds roaming the beaches as late as June 5 looking for food. By then, it is almost too late to reach the Arctic in time to breed successfully.

Timing is all. If winter lingers and the bay doesn't warm by mid-May, horseshoe crabs may not be spawning when birds arrive. In addition, difficulties the birds face in one stopover can reverberate along the entire migration route. If knots tarry in South America, if food shortages in Tierra del Fuego or San Antonio Oeste delay their departure, there may not be enough time in Delaware Bay to gain the necessary weight. In Delaware Bay there is little slack, and the numbers of shorebirds are still not high enough to absorb natural perturbations here, let alone ones we may add. Amanda Dey describes it this way: "The capacity of diminished shorebird populations to absorb bad times is slim. The longer the stopover remains in this condition, the more this becomes a game of Russian roulette."

The numbers of knots in the bay may have stabilized, but the population is far from recovered. The numbers offer hope that restoration is possible. Yet, like the shorebirds themselves, who've come so far to reach Delaware Bay, and who still have so many miles to travel, the journey of restoration has a very long road ahead. From Niles's perspective, the fight to bring back horseshoe crabs, ongoing for 15 years, hasn't been won yet. David

Mizrahi from the New Jersey Audubon Society, who monitors semipalmated sandpipers migrating through Delaware Bay, agrees. "We're not there yet," he tells me in a phone call. "The eggs still aren't what they used to be. There are enough now to support the low numbers of birds flying through the bay, but not yet enough to rebuild diminished populations." Later, he put it another way: "Only increasing shorebird populations in Delaware Bay coupled with improving energetic condition [more and fatter birds] would signal improving horseshoe crab populations." Still-diminished numbers of horseshoe crabs may support still-diminished numbers of shorebirds, but there aren't yet enough horseshoe crabs to sustain a restored population of red knots. Shorebirds are still on the edge. Egg densities still aren't what they once were, and knot counts are still down by two-thirds. Ruddy turnstones are still in trouble and so are semipalmated sandpipers.

Restrictions in the horseshoe crab fishery allow fishermen to continue taking horseshoe crabs for bait while theoretically leaving enough horseshoe crabs in the sea to enable the population to rebuild. After more than a decade of regulation, however, horseshoe crabs are continuing their downward trajectory in New York and New England; the yearly take there is no longer sustainable. Horseshoe crab population growth in Delaware Bay, according to the U.S. Fish and Wildlife Service, has "stagnated." The horseshoe crabs have yet to rebound. The relative abundance of young horseshoe crabs in the bay, harbinger of a population rise down the road, increased in 2009, but then dropped again. The number of adult female horseshoe crabs has yet to increase. Scientists estimate that Delaware Bay could support 14 million female horseshoe crabs; today there are only 4 million.

Perhaps not enough time has passed. Female horseshoe crabs mature in eight to 10 years, but numbers of immature crabs should be rising more rapidly by now. The quotas may be too generous to restore horseshoe crabs and red knots to early 1980s levels of abundance. Under the current plan, regulators may be setting their sights too low: it will take decades, maybe as many as 60 years, before the number of female horseshoe crabs rises to half what the bay can support. Furthermore, the rules, designed to help red knots, don't consider the needs of declining ruddy turnstones and semipalmated sandpipers. When horseshoe crabs are so necessary to human health and to the health of shorebirds, perhaps it is time to consider whether it is necessary to take so many horseshoe crabs for bait, whether more

horseshoe crabs are killed than the quota suggests, and why the population
is not rebounding.

On rainy nights in autumn, at the full and new moon tides, eels,
normally creatures of darkness, emerge from the muddy bottoms of ponds
to slip through streams and rivers to the ocean and out to the Atlantic's
distant Sargasso Sea to spawn. In the spring, young glass eels reverse the
journey, making their way up estuaries, following the scent of freshwater.
Like red knots, they are great migrators, swimming as far inland as the tributaries of the Mississippi and the edges of Lake Ontario, but unlike the
birds, they make their long round-trip journey only once, dying after they
spawn. They live as far north as Newfoundland, and as far south as French
Guiana. Once they were prolific in North America. In the 1600s, Jesuit
missionaries on New York's Lake Onondaga witnessed 1,000 speared in a
single night. Colonists at Plymouth took eels by the hogshead. One fish out
of every four in East Coast streams was an eel. New Jersey fishermen caught
eel in oak pots baited with horseshoe crabs, preferably female, "chopped in
half or quarters."

Fishermen still take horseshoe crab to catch eel. The eels themselves
are used as bait by striped bass fishermen, or they are shipped overseas and
sold as delicacies, fried, gelled, or smoked. Between 1981 and 2010, as many
as 6 million pounds of eel were exported annually from the United States.
In 2012, glass eels exported from Maine sold for $2,000 a pound, with indi-
viduals earning $150,000 in a fishery worth $39 billion. In 1969, 21 million
pounds of eel lived in the nation's freshwater streams and ponds. Today, 4
million pounds remain. Overfishing, hydroelectric turbines and dams, and
parasites led to their decline. Eels, declared the ASMFC in May 2012, are
depleted.

Fishermen also take horseshoe crabs to catch whelk, large snails living
in the sand and mud of coastal waters, popularly called conch. Occasionally,
I see their eight- or nine-inch-long whorled shells washed up on the beach.
More often, I find their egg cases, long strands of pale, translucent discs,
necklaces holding hundreds of eggs. (In Italian American communities,
whelk, or scungilli, can be on the menu for Christmas Eve, at the Feast of
the Seven Fishes. A recipe for scungilli in a sauce of tomatoes, cloves, and
red wine, along with an essay by Tony Soprano's psychiatrist Jennifer Melfi,

appears in the chapter "Rage, Guilt, Loneliness and Food," in *The Sopranos Family Cookbook*.) Most whelk caught in the United States are exported to Asia and Europe, where they are served fried, curried, and in chowder. In 1999, Virginia fishermen took 1.4 million horseshoe crabs to catch whelk.

Along the eastern seaboard, large catches of whelk increased 62 percent from 2005 to 2010 and then dropped, characteristic of a potentially insufficiently regulated boom-and-bust fishery. In Massachusetts waters, very few mature female whelk remain. As of 2014, the ASMFC had neither taken the measure of whelk in a stock assessment nor set quotas for the fishery.

Eel are depleted, whelk are falling off, and horseshoe crab populations are stagnating in Delaware Bay and declining in New England. Where is memory? The nineteenth-century cycle of massive extraction and decimation repeated itself as rising demand for eel and whelk depleted horseshoe crabs once again. In the blink of an eye, eel and whelk may come—and go. They slip away in a broader context of loss. Delaware River shad, once the most prolific fishery in any Atlantic river, now at an all-time low, have yet to recover. River herring, depleted, are still declining. Delaware River sturgeon, once the largest population in the United States, are now on the U.S. Fish and Wildlife Service's endangered species list, with fewer than 300 adults spawning each year. All these fish are managed by the ASMFC. We look back, askance, at market gunners who killed thousands of shorebirds and at fishermen who turned millions of horseshoe crabs into fertilizer, but one day our children may ask how we, with more awareness, took one diminished species as bait to catch other species, themselves diminished.

It may be shortsighted and imprudent to take so many eel and whelk: it is unnecessary to kill so many horseshoe crabs to do so. A waterman from Bowers Beach, Delaware, Frank "Thumper" Eicherly IV, stretched his horseshoe crab bait, protecting it in a net bag that hung in his traps, beyond the reach of ravenous whelk. He caught the same number of whelk with half the number of horseshoe crabs. The Virginia Institute of Marine Science and Glenn Gauvry from the nonprofit Ecological Research and Development Group made and distributed 14,000 bait bags to whelk fishermen, free.

What Eicherly and Gauvry began, Nancy Targett, dean of the University of Delaware College of Earth, Ocean and Environment, continued. Originally, Targett, working with Dupont, tried to identify and isolate the special scent of horseshoe crab, a stench to the Camp children growing up near the fertilizer factory but irresistible to whelk and eel. Fifteen years later, the horseshoe crab's unique essence still eluded her, but in the spring of 2013, she cooked up an alternative, a seaweed gel spiked with bits of horseshoe crab. Substituting a little Asian shore crab cut down the need for horseshoe crab even further. Her bait saves horseshoe crabs and helps get rid of Asian shore crab, pests that arrived in the United States in 1988, making landfall in Cape May—hot spot for migrating birds. The birds fly in: the crabs came by boat, in ballast water through the Panama Canal.

They quickly worked their way up and down the East Coast, pushing other shellfish aside. Opportunistic omnivores, they compete with blue crab and lobster for food. Targett's bait without Asian shore crab uses one-eighth of a horseshoe crab. Supplemented with the invasive species, her bait uses one-twelfth of a horseshoe crab. Delaware Sea Grant provides a recipe, which fishermen can whip up in a blender, and LaMonica Fine Foods, in Millville, New Jersey, purveyor of surf clams, quahogs, and scungilli, manufactures and sells it under the name Eco-Bait. If horseshoe crab bait fishermen were to use Targett's bait, the take of horseshoe crab for bait could be reduced by 75 to 90 percent. Thanks to Eicherly, Gauvry, and Targett, there's no need to take so many horseshoe crabs.

There is no horseshoe crab bait industry in South Carolina. When his company was young and getting established, Jim Cooper wrote me he was "shocked" to learn that fishermen delivered "truckloads" of horseshoe crabs to a tractor trailer in Beaufort, where baiters paid them "a quarter per crab." Working with the South Carolina Department of Natural Resources, he drafted legislation to end the South Carolina bait fishery. Aided by a member of his church and his wife, Frances, who, he said, "guided the legislative process from Charleston," the legislation passed in 1991. The ban was foresighted. The biggest horseshoe crabs along the East Coast are in South Carolina: averaging almost one foot across, their size has remained constant over the years, suggesting a stable population.

Bait made with seaweed gel flavored with Asian shore crab and horseshoe crab would alleviate another threat to horseshoe crabs, one that regulators hadn't anticipated and that, had it gone unchecked, might have had potentially serious consequences. In a candlelit London crypt, the pop-up

Café de Mort served, to guests who'd signed liability waivers, a dinner deadly if not properly prepared. An ambulance waited nearby. The menu for this event, which took place in 2013, included snake wine (a coiled cobra inside the bottle) and pufferfish, perhaps the most poisonous fish in the sea, served by a chef who trained for 10 years before receiving his license to prepare them. In Japan, pufferfish are a delicacy. In some species, the entire fish is lethal. In others, the potent neurotoxin, tetrodotoxin (TTX), accumulates in the liver and intestines, requiring highly trained chefs to slice it without contaminating the flesh. TTX is 10,000 times more potent than cyanide: one forkful of TTX-laced pufferfish can kill a person in a few hours. Their import into the United States is highly restricted.

Other marine animals harbor TTX, including an Asian horseshoe crab whose toxic eggs have hospitalized dozens of people in Thailand and Cambodia. Consuming half a cup of eggs can be fatal. Given the toxicity of TTX, importing Asian horseshoe crab into the United States threatens American horseshoe crabs, shorebirds eating crab eggs, whelk and eel eating crab, and humans eating whelk and eel. When Asian horseshoe crabs began appearing in the U.S. bait market (2,000 in 2011 and 4,000 in 2012), the ASMFC, at the urging of the IUCN Horseshoe Crab Specialist Group, banned them.

Invasive species can wreak havoc in ways unanticipated and difficult to rectify. A parasitic flatworm native to Japanese eels hitched a ride into Europe, infecting eels there, and then crossed the Atlantic with unfortunate result. *Anguillicola crassus* first appeared in a wild eel in Winyah Bay, South Carolina, not far from the horseshoe crab spawning beaches in Bulls Bay. Highly pathogenic, the worm feeds on eel blood, causing hemorrhaging and thickening of their swim bladders, impeding their ability to swim, and increasing odds that infected eels will arrive late to their spawning grounds without sufficient energy reserves to produce high-quality eggs. Scientists believe that, for eels, this parasite is a "serious threat for the overall reproductive success."

The parasite spread rapidly across river basins and estuaries, as humans caught eel in one estuary and sold it for use in another as bait. I remember peering into a tank of live eel at a local bait store, brought in from Maryland to be chopped up for Massachusetts striped bass fishermen. Nobody thought anything of it. Initially, the parasite wasn't a great concern since it didn't appear to hurt Japanese eels. Now, though, for American eels, a species already under duress, it makes their restoration that much more difficult.

American horseshoe crabs aren't immune from this kind of accident. Asian horseshoe crabs harbor parasitic flatworms that, like those of eels, can easily hatch and survive in the water, even if the crab is killed for bait. The gills of a fish are near its head; the gills of a horseshoe crab—hundreds of thin membranes that can be flipped like pages in a book—are near its tail. Just as the Asian *Anguillicola crassus* didn't harm Japanese eels but severely impaired American eels, infestations of Asian flatworms into the gills of American horseshoe crabs may inhibit their ability to draw oxygen from the water.

Counting horseshoe crabs continues. The numbers don't always add up. It's time to understand why horseshoe crab numbers continue to decline in New England, and why they are not rebounding in Delaware Bay. The bait quota may be set too high, especially if horseshoe crabs are being killed that aren't counted in the quota: Niles and Dey describe the system governing horseshoe crab quotas as "leaky." In a letter to the U.S. Fish and Wildlife Service, New Jersey Audubon, the American Littoral Society, the Delaware Riverkeeper Network, and the Conserve Wildlife Foundation of New Jersey all expressed concern that the ASMFC quota hasn't taken into account illegally caught horseshoe crab and horseshoe crab caught and discarded by commercial fishermen targeting other fish or shellfish.

How the biomedical take of horseshoe crabs affects the status of the population isn't well understood. The data haven't been part of the ASMFC horseshoe crab stock assessments. Technical committees of the ASMFC would like these numbers included so that "an accurate portrayal of [horseshoe crab] removals is occurring." They would also like this information broken out by region, enabling a clearer understanding of the potential for the growing biomedical catch to contribute to the continuing decline of horseshoe crabs in New England or the slow recovery in Delaware Bay. Because there are only one or two places in each region where horseshoe crabs are bled, the companies consider this information confidential. The ASMFC and the biomedical industry are looking at ways to incorporate regional data in the stock assessments.

To assure the long-term health of horseshoe crabs, it may be necessary for the ASMFC to evaluate not only the biomedical catch of horseshoe crabs by region but also the number of biomedical horseshoe crabs that die every year. The ASMFC Horseshoe Crab Stock Assessment Committee considers exclusion of these data an "oversight" in need of remedy, since this information "may account for a significant portion of the annual exploitation of horseshoe crabs in the Delaware Bay region." Advisors to the ASMFC—the horseshoe crab technical committee—are concerned that the number of horseshoe crabs that die in the biomedical industry "will eclipse management efforts focused on the bait fishery."

In the past, blue bloods taken for bleeding constituted a small percentage of the horseshoe crab catch. They were returned to the water, and the number that died compared to the 100 percent death rate in the bait industry was inconsequential, but a once-small industry has rapidly grown: the number of horseshoe crabs taken for biomedical use climbed by more than 300 percent between 1989 and 2012 and, according to ASMFC advisors, is now "essentially equal" to the number taken for bait.

The ASMFC has set a threshold—57,500—for the number of horse- shoe crabs that could be killed or die in the biomedical industry each year, a threshold that the industry has exceeded every year since 2007, sometimes by as much as 40 percent. In setting the threshold, the ASMFC assumed that each year 15 percent of biomedical horseshoe crabs would die, either when they were bled or when they were returned to the sea. The industry claims that the number of deaths is lower than these estimates. Four recent studies offer a higher number, suggesting that as many as 20 to 30 percent of bled female horseshoe crabs may die. These findings, highlighting another potential "leak" in the system, could substantially increase horseshoe crab losses in the biomedical industry each year to numbers that may equal or exceed the bait take in a number of states.

On dark, moonlit beaches lined with spawning horseshoe crabs, I've come to expect four, five, six, sometimes 10 or 11 males clustered around a single female. These larger male crowds may be a relatively recent phenom- enon, the result, scientists suspect, of biomedical and bait fishermen selecting out the larger females. Historical accounts suggest such skewed ratios are not the norm. Richard Rathbun, writing for the U.S. Fish Commission in 1884, included an account of one observer that horseshoe crabs came ashore to spawn "in pairs." Rathbun added that "it is not an uncommon thing for the female, as she crawls up the beach, to be accompanied by two, three, and even as many as six males. . . . As a rule, however, each female brings with her only a single crab." Hugh M. Smith, describing the king crab fishery in Delaware Bay in 1889, wrote that although female crabs coming up onto sandy beaches to spawn "sometimes" were accompanied by "two or more males," usually they "seek the sandy shores in pairs."

In Pleasant Bay, Cape Cod, where bait and biomedical fishermen have taken horseshoe crabs for over 25 years, the proportion of spawning male and female horseshoe crabs has become increasingly and alarmingly skewed. The ratio, once two or three males for every female, increased to one female for every five or six males in 2000–2, and then to one female for every eight or nine males in 2008–9. These "extreme" sex ratios, as described by one scientist, included clusters of 12 males for every female and even a cluster of 30 males for a single female. In Cape Cod's Nauset estuary and in the Monomoy National Wildlife Refuge, where there is no longer a horseshoe crab fishery, only one or two male crabs come ashore for every female. Today, the Delaware Bay spawning survey shows increasing ratios of males, up to three to five male crabs for every female.

Horseshoe crabs taken to make LAL may suffer sublethal effects of handling and bleeding that inhibit spawning. Fishermen collect horseshoe crabs by hand or rake where they are spawning, as Jerry Gault does in South Carolina. They also dredge or trawl for horseshoe crabs farther offshore, scraping the animals from the seafloor and then dropping them onto the decks of boats. Scientists examining these horseshoe crabs have observed "traumatic injuries," including fractures in their shells, often with the crab's digestive and reproductive organs visible, and "stab-like wounds," where one crab's tail punctured another crab's gills when they were struck by dredges or piled in a boat.

Horseshoe crabs can become stressed when they are taken from the sea for bleeding: they can be piled on boat decks or stacked in bins while fishermen are working; crowded in holding pens or ponds; and subjected to heat and dehydration when they are towed in skiffs or driven in non–air-conditioned trucks to be bled. Horseshoe crabs can be kept out of the water for the better part of a day, and up to 72 hours. Furthermore, bleeding a horseshoe crab is not the same as one of us going to the Red Cross to donate blood. Human blood donors usually give about one pint, 10 percent of their blood; up to 30 to 40 percent of a female horseshoe crab's blood may be extracted.

Bled crabs tagged with radio transmitters returning to the waters of Pleasant Bay, Cape Cod, wandered aimlessly along the seafloor, disoriented. Bled horseshoe crabs from Great Bay, New Hampshire, became lethargic two weeks after they were bled, and failed to respond to tidal rhythms that call horseshoe crabs to spawn. Six weeks after bleeding, only 60 percent of their hemocyanin, the protein that makes up 90 percent of their blood and that carries oxygen throughout their circulatory system, had been restored. It may take six months for their blood supply to be fully replenished. Is culling spawning horseshoe crab by hand more or less protective of horseshoe crabs than dredging or trawling them? How many surviving horseshoe crabs spawn? That number needs counting, but ASMFC stock assessments and management plan reviews don't seem to contain that kind of information.

The system for managing horseshoe crabs contains another "leak." In 2001, to protect horseshoe crabs concentrating at the mouth of Delaware Bay from trawlers and dredges, the ASMFC asked the National Marine Fisheries Service to establish a horseshoe crab reserve at the mouth of Delaware Bay, where "taking of horseshoe crabs for any purpose, including biomedical would be prohibited." The reserve was designed to protect "older juvenile and newly mature females." By the time the 1,500-square-mile Carl N. Shuster, Jr. Horseshoe Crab Reserve was established, the biomedical industry had obtained an exemption, permitting it to remove 10,000 crabs a year from the reserve. Is the reserve large enough to provide sanctuary for horseshoe crabs spawning in Delaware Bay? Is its ability to protect crabs diminished when horseshoe crabs are removed for biomedical use? When horseshoe crabs are failing to rebound, we may need to reexamine whether the reserve is functioning as it was intended.

The ASMFC is working with the biomedical industry to establish voluntary best-management practices. These could include prohibiting a preferential take of female horseshoe crabs, carrying them by their shells rather than their tails, keeping them moist and out of the blistering heat, transporting them in air-conditioned or refrigerated trucks, tagging them to ensure they are not bled twice in a summer, monitoring oxygen levels throughout the season in ponds or pens where they are held, and returning them gently to the sea within 24 hours. Can these practices, many of which are theoretically already in place, further reduce biomedical mortality, reduce the sublethal effects of bleeding, and ensure that female horseshoe crabs continue spawning? Are they enforceable? Industry standards and practices, open to scrutiny and regularly audited, would be stronger than voluntary best-management practices.

In the end, the numbers that count are those indicating whether depleted horseshoe crab populations are being rebuilt, a determination made more difficult in Delaware Bay by the cancellation in 2013 of the annual Virginia Tech horseshoe crab trawl survey. Right now, in New England and Delaware Bay, those numbers suggest that to protect human public health and the health of shorebirds, horseshoe crab mortality needs to drop. A coastwide moratorium in the bait industry and switch to the alternative bait would achieve that, as would tighter regulation of the biomedical industry. Understanding the full extent of horseshoe crab mortality and reducing it is imperative: biomedical demand for horseshoe crabs is rising.

Glenn Gauvry lives in Little Creek, Delaware, in a 220-year-old house that, little by little, piece by piece, he is lovingly restoring. The house, with its low ceilings and wide board flooring, had been falling apart, abandoned, and new owners, confronted with the widespread deterioration and monumental repairs, came and went in rapid succession. Just as he is carefully, painstakingly restoring his house, so he is carefully, lovingly caring for horseshoe crabs. A big focus of his work is across the world, in estuaries along the coasts of Taiwan, China, Indonesia, and the Philippines; at the mouth of the Ganges River; and at the edges of the Bay of Bengal. Asian horseshoe crabs, once prolific, are disappearing, too many taken for consumption and biomedical use, and too few of their homes safeguarded. In Japan, they are an endangered species. At an international conference on the science and conservation of Asian horseshoe crabs, the presentations, one after another, spoke of ineffective regulation and illegal catches; of coastal and watershed development destroying the crabs' homes and extirpating their populations; of difficulties, once crab are gone, of restoring degraded habitat; of abundance that has faded away; of profound loss and the difficulty of U.S. crab meeting rising demand.

The already colossal world market for pharmaceuticals and medical devices is escalating. By 2016, world spending on drugs, increasing at a rate of 3 to 6 percent per year, will approach a staggering $1.2 trillion, fueled by emerging markets in China, India, and Brazil, where the pharmaceutical industry is projected to grow at annual rates of 12 to 15 percent per year by 2016. Global demand for vaccines alone is forecasted to reach $56 billion by 2016. Analysts project the world market in medical devices will grow 6 percent per year, reaching $302 billion by 2017.

For drugs, medical devices, and vaccines that require testing for life-threatening endotoxins, rising demand coupled with a decline in Asian horseshoe crabs to "critical levels" may lead, Gauvry believes, to a shortage of horseshoe crabs and, when Asian horseshoe crabs can't supply growing demand, to biomedical companies turning to horseshoe crabs from the United States. There is no limit in the United States on the number of horseshoe crabs companies may take to bleed, but the biomedical industry has already exceeded the threshold for how many may die every year in the process. It may be difficult to increase the take without further violating the threshold. For Gauvry, undaunted by the challenges, saving the horseshoe crab is an ethical imperative that requires economic solutions. He, researcher Mark Botton (a biologist and expert on horseshoe crabs), and many colleagues have successfully petitioned the IUCN to begin the long process of determining whether Asian horseshoe crabs belong on the Red List of Threatened Species.

At the same conference where Gauvry and other scientists described the critical decline in Asian horseshoe crabs, John Dubczak from Charles River Laboratories presented one economic solution: using considerably less LAL in endotoxin tests. In his presentation, Dubczak described the company's FDA-approved cartridge, the one I saw in use at Mass General. He said it uses 95 percent less LAL. Researchers at the conference also presented a genetically engineered alternative to LAL.

Many commonly used synthetic pharmaceuticals originated from wild plants and animals. Jesuit missionaries in Peru observed that Quechua Indians alleviated shivering—induced by the long hours in cold Spanish mines—by chewing on chinchona bark. The priests took the bark to Europe, where it treated fever and malaria. The newer malaria drugs chloroquine and mefloquine contain synthetic analogs to the bark's medicinal quinine. Eventually, chloroquine-resistant malaria and mefloquine-induced anxiety, vivid dreams, and hallucinations necessitated another approach. During the 1960s, when drug-resistant malaria was killing North Vietnamese fighters, scientists working secretly for Mao Zedong developed what would become a next-generation malaria drug, derived from another plant, aromatic sweet wormwood. After 12 years of research, researchers made a semisynthetic version—artemisinin.

Half the drugs prescribed in the United States today originated in the healing properties of wild plants and animals: aspirin, synthesized from salicylic acid, a compound in willow leaves and bark; the blood thinner warfarin, synthesized from coumarin, discovered when cows eating spoiled clover sometimes hemorrhaged and bled to death; heparin, another widely used anticoagulant, synthesized from pig intestines; the breast cancer drug taxotere, synthesized from taxol in the bark of Pacific yew trees. Synthetic drugs inspired by nature are everywhere, although as synthetic drugs come to dominate our medicines, their association with the natural world from which they came fades. Scientists might not have developed the common antibiotic amoxicillin without Alexander Fleming, who in 1928 returned from vacation to his lab in London's St. Mary's Hospital to find, in an accidentally contaminated petri dish, Staph bacteria stopped in their tracks by a blue-green mold, *Penicillium*. The road to Lipitor, a synthetic statin and the world's best-selling drug, began in the lab of Akira Endo, who located cholesterol-inhibiting compounds in mold isolated from rice he'd purchased in a store in Kyoto.

An Asian horseshoe crab, *Carcinoscorpius rotundicauda*, thrives in bacteria-laden estuaries in Singapore. Smaller than its American counterpart, it contains less blood but a more potent endotoxin detector. Jeak Ling Ding, Bow Ho, and their colleagues at the National University of Singapore undertook the long and difficult work of isolating and genetically engineering Factor C, the protein that triggers clotting in horseshoe crab blood. Eventually, they cloned the protein, then recloned the synthetic DNA in yeast and in cells derived from monkey kidneys and insects.

When crabs were spawning in Delaware Bay, and Richard Weber and I were breathing gnats on egg-strewn beaches, another animal was also laying its eggs, 400 miles away, in pastures in western New York, between the Finger Lakes and Lakes Erie and Ontario. Storms had blown in adult moths, whose larvae had hatched and grown to one and a half inches long. The ravenous caterpillars ate their way through crops of alfalfa and corn, wheat and rye and, like Sherman marching to the sea, devoured everything in their path. When they were finished, the U.S. Department of Agriculture declared 13 counties in New York a natural disaster. Armyworms may be the scourge of farmers, but they are of interest to drug companies making synthetic bacterial endotoxin tests. The National University of Singapore licensed its patent for recombinant Factor C (rFC) to Lonza. The company makes LAL and manufactures recombinant rFC, inserting its DNA code into a cell line derived from armyworms.

Karen Zink McCullough, editor of a 400-page handbook on endotoxin testing and consultant to the pharmaceutical industry, conducted endotoxin tests for Beecham Laboratories, well before it was absorbed into SmithKleinBeecham, before Wellcome merged with Glaxo, and before they all merged into the giant GlaxoSmithKlein. "I've been in this almost as long as Jim Cooper," she says, remembering when she first began using the toxin detector made from the blood of horseshoe crabs. "We had to send our injectable drugs to the FDA for rabbit testing," McCullough recalls. The company's rabbit test was negative for one lot; the FDA's test was positive. "My boss showed me Jim Cooper's LAL paper and said, 'Why don't you try some of that stuff?' . . . That was the beginning. LAL has served us well for more than 30 years, but now there are new models for testing endotoxin, harbingers of new things on their way."

The genetically engineered protein from horseshoe crab blood is one. In another new test, developed in Europe, human or cultured white blood cells detect fever-inducing Gram-negative bacteria and other fever-inducing bacteria, viruses, and parasites. In Singapore, Ding, Ho, and others, continuing to tease out the ingenious ways horseshoe crab blood snares endotoxins, identified an amino acid that, if placed in a sticky resin, traps endotoxin in water used to manufacture drugs and medical devices.

Many years elapsed before the FDA and industry were convinced of LAL's superiority to rabbit tests. Another long and convoluted road to acceptance may lie ahead for the genetically engineered alternative to LAL. A blood product, LAL is licensed by the FDA. The recombinant doesn't easily fit into existing FDA jurisdictions. It's not a drug, medical device, or diagnostic test for patients, and it doesn't emit radiation. Currently, the agency doesn't see the need to license it. "rFC doesn't seem to have a real home there," says McCullough. She told me that other quality-control tests used by drug companies that are chemical or physical in nature—as opposed to biological—such as tests for pH, clarity, or residual solvents, don't require FDA licensing. Unless there's a compelling need, there may be little advantage to pharmaceutical companies in switching from a gold-standard FDA approved test to one that doesn't require FDA approval.

Pharmacopeia literally means "preparing drugs." Standards and procedures for quality-control testing of drugs are found in the United States Pharmacopeia (USP). The USP, the nation's official reference for prescription and nonprescription drugs, contains standards for their preparation and use. Scientific experts set criteria for the pharmacopeia, which drug manufacturers must follow, for the formulation, quality, purity, and strength of drugs. Regulators in the United States, Europe, and Japan "harmonize" their standards for bacterial endotoxin testing. All three references now contain identical testing methods and standards, reducing the regulatory burden on global pharmaceutical companies but adding additional layers of approval to the process of considering new tests for inclusion. The 2014 edition contains standards for rabbit tests and for LAL made from horseshoe crab blood, but not for the newer synthetic test.

Lonza has published one study examining the equivalency of its LAL and rFC, building a case to include the recombinant in the USP. In the early days of LAL, one major manufacturer of IV solutions conducted over 140,000 LAL tests and 28,000 rabbit tests on its products, confirming that LAL was more sensitive. Cooper, collaborating with the FDA, ran 155 tests on radiopharmaceuticals and other drugs. The FDA carried out its own research as well. There was safety in these numbers: tests revealed properties of drugs that could sometimes interfere with LAL, altering results and requiring further research and refining. Fewer tests will probably be required for rFC to prove its equivalence to LAL. The USP committee undertaking the last revision of the standards for endotoxin testing noted that an "important reason for revision of the harmonized test may be advances in technology." If or when standards for endotoxin testing include genetically engineered LAL, the decision will be made jointly by the United States, Japan, and Europe.

In 2012, the FDA began allowing companies to use rFC if they demonstrated, according to USP standards, that it showed equivalent or better results than tests already in the USP. "It may be hard," McCullough says, "to turn the *Titanic.*" LAL is a known, reliable test licensed by the FDA. Companies using it have little incentive to spend time and money validating the newer, less familiar, and less proven rFC, unless they are developing drugs whose properties aren't compatible with rabbit and LAL testing (as rabbit testing proved unsuitable for radiopharmaceuticals 40 years ago) or unless the time comes when there aren't enough horseshoe crab to meet demand. Allen Burgenson, manager of regulatory affairs at Lonza, is taking the long view. "Some twenty years elapsed between the time scientists described the clotting mechanism in horseshoe crab blood and large numbers of pharmaceutical companies replaced the rabbit test with LAL," he says. "We're only 10 years into rFC."

In the meantime, horseshoe crabs are still spawning. Red knots are still flying in to eat their eggs. I felt the beach calling. Although Delaware Bay is the best-known and largest spring stopover for knots, I wanted to visit a few others. And since South Carolina had so many horseshoe crabs, I wondered what shorebirds might be there, eating eggs.

Eight

LOWCOUNTRY

South Carolina and Other Tidelands

I am drawn again and again to South Carolina, to the seeming time-
lessness of a coast whose soft contours the sea continuously redraws
and to an expanse of porous salt marsh threaded by tidal creeks, so
many I could spend all my days paddling among them and never
really know them. Dear friends Ellen Solomon and Richard Wyndham live in
a home set back from a road lined with live oaks, their yard imbued with the
heady aroma of jasmine, mock orange, and gardenia. From there, we can't
hear or even see the sea, but saltwater has known their land: an 18-foot storm
surge from Hurricane Hugo rolled six miles over beach and salt marsh and
through the woods, filling the yard with water over six feet deep.

They live at the edge of the Cape Romain National Wildlife Refuge—
65,000 acres of salt marsh, sandy beach, and open water extending 22 miles
along the South Carolina coast. Wyndham loves to pore over old maps,
resurrecting creek names long forgotten and charting the coming and going
of sand. Each time I visit, we go out in his johnboat, down creeks that wind
through the marshes, one after another—Home Creek to the Intracoastal
Waterway to DuPre, Casino, Skrine, and Congaree Boat creeks, then on to
Tower and Horsehead creeks, Romain Harbor, Slack Reach, and the Romain
River, before emptying out behind the refuge's slender barrier islands.
Another route goes out into the broad open water of Bulls Bay.

Loggerhead turtles nest on the beaches of Cape Romain, some 2,000, the largest loggerhead nesting site north of Cape Canaveral. Fertile Cape Romain, with an abundance of shrimp and acre after acre of protected creek, is an ideal nursery for shark. Finetooth, blacktip, sandbar, Atlantic sharpnose, scalloped hammerhead, and a newly recognized species—the Carolina hammerhead—raise their young here. Scientists caught a six-foot-long scalloped hammerhead in Five Fathom Creek, a creek I have traveled several times. Seventy miles seaward of the refuge tiger sharks give birth—in one of the highest concentrations of tiger sharks and newborns in the western Atlantic. Biologist Felicia Sanders has seen sharks in the refuge, as has Wyndham: three or four just off the beach, rolling under the stern of his boat in only a few feet of water, where he'd been wading as he pulled the bow ashore. I've heard stories of sharks brushing against people standing in waist-deep water in the refuge, but if sharks swim beneath the boat, I neither see nor sense them.

Horseshoe crabs, unpalatable to some animals, are important prey for refuge inhabitants. Sea turtles, fish, and horseshoe crab shells fill the stomachs of Carolina tiger sharks. Hammerhead sharks regularly feed on horseshoe crabs and, in all likelihood, sandbar and blacktip sharks as well. Horseshoe crabs are the preferred food of loggerhead sea turtles, listed as threatened under the Endangered Species Act. The turtles come into estuaries when horseshoe crabs are plentiful, easily scooping out eggs, gills, and legs with their thick beaks and large jaws.

I'd gone to Delaware Bay to witness the life-sustaining connection between shorebirds and horseshoe crabs. I'd come to South Carolina to accompany fishermen catching horseshoe crabs, to follow horseshoe crabs from the sea, and to learn how their blood is turned into a toxin-detecting test that touches my own life, my family's lives, and the lives of everyone I know, forging a connection between humans and these ancient animals that is as strong as their tie to shorebirds. I want to see shorebirds in South Carolina. The Cape Romain National Wildlife Refuge has recorded 293 species of birds, including many seabirds and shorebirds—nesting skimmers and brown pelicans; wintering oystercatchers; royal, least, sandwich, gull-billed, Forster's, and common terns; migrating red knots, dunlin, marbled godwit, ruddy turnstone, whimbrel, willet, and short-billed dowitchers whose bills, confusingly, are occasionally longer than their long-billed relations. Walking the long

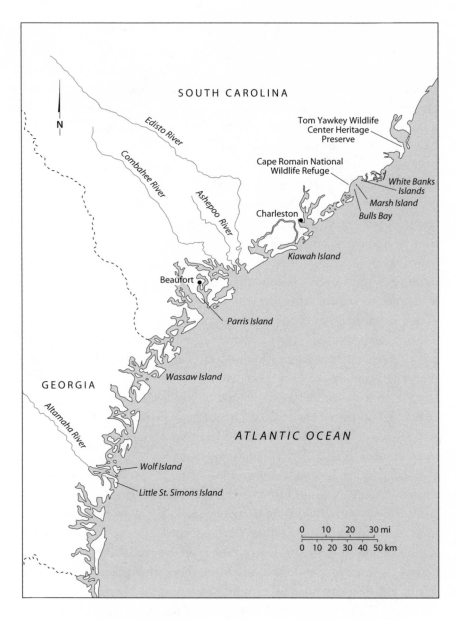

South Carolina and Georgia (map by Bill Nelson).

slow curve of beach, stepping around giant cannonball jellfyfish washed
ashore, I come upon my first long-billed curlew, its slender bill, a full six
inches long, curved like the crescent of a barely new moon.

Only about 140,000 remain of the long-billed curlew, North America's
largest shorebird. I wonder how many more times I will see it. Audubon,
describing long-billed curlews flying in at sunset to roost on the beaches
near Charleston, wrote: "The flocks enlarge ... as they proceed, and in
the course of an hour or so the number of birds that collect in the place
selected for their nightly retreat sometimes amounts to several thousand
birds." Today, the Cape Romain National Wildlife Refuge is the only place in
South Carolina where long-billed curlews are still regularly seen, but even
here they are disappearing. In 2001, Sanders regularly saw between eight
and 12 birds in the refuge. These days, an observer in the refuge has seen
only two.

Sanders manages shorebird and seabird protection and conservation
for the State of South Carolina. She and her husband eat from the land,
catching fish and shrimp, hunting and butchering deer, turkey, and feral
hog. When I first met her, she was preparing to haul a deer from the woods.
She will go anywhere, anytime, to check on the birds she loves, once kayaking
across the refuge, through the marshes, down the Romain River, and out to
the beach to watch birds, an 11-hour round trip. I hope to see a few red
knots when I am in South Carolina, but at the time South Carolina didn't
figure prominently in the story of red knot migration and conservation. I
am lucky to be in South Carolina as Sanders and her colleagues are
confirming the state as a critical stop on the knot migration route. Their
findings are redefining our understanding of where red knots and other
shorebirds find home, and what protecting them entails.

Ornithologists were watching knots in South Carolina more than
100 years ago. Arthur Trezevant Wayne, an aristocrat born during the
turmoil of the Civil War, worked briefly and miserably for a Charleston
cotton merchant before going on to become a prominent ornithologist,
killer of ivory-billed woodpeckers, and author of *Birds of South Carolina*.
He watched birds whenever he could, walking and rowing miles to find
them. In May 1895, he saw 3,000 knots on a beach outside Charleston and
noted that knots had a history of abundance on Bulls Island, now part of
the refuge. In 1949, Alexander Sprunt and E. Burnham Chamberlain,

authors of *South Carolina Bird Life,* confirmed the knot as a winter resident in South Carolina. To them, the bird connoted "an untrammeled wildness and freedom that is equaled by few and surpassed by none."

Sanders takes me out through Five Fathom Creek to shrubby Marsh Island and the nearby White Banks Islands to look at birds. The water is shallow, between one and three feet deep. Occasionally we bump bottom. Barely rising above the waters of Bulls Bay, low-lying Marsh Island and White Banks Islands are blanketed with thousands of nesting and roosting birds: black skimmers, white pelicans, terns, and laughing gulls. We see an oystercatcher nest in the wrack line. We don't see any knots, but as many as 3,000 have been known to roost here, and another 600 nearby. Sanders, watching knots in late May in Cape Romain and on Harbor Island, near Beaufort, saw orange and red flags from Argentina and Chile, but rarely lime green. She realized that some knots might refuel in South Carolina, with its abundance of horseshoe crabs and sheltered inlets, and then go straight to the Arctic, bypassing Delaware Bay. Brian Harrington, who'd looked for knots all along the eastern seaboard, realized this as well: observers in Delaware Bay see very few birds also seen in South Carolina or Georgia.

Sanders and biologist Janet Thibault conducted spring and fall surveys south of Charleston, and on Kiawah Island, they found 8,000 knots on one day in March 2012, confirming Brian Harrington's observation of Kiawah's importance to knots at least a decade earlier. Not all South Carolina's red knots are red. In June 2010, residents on Kiawah Island alerted biologist Aaron Given to an unusual sighting, an albino knot, all white except for a blush of red along its throat and chest. Its feathers had lost their pigmentation, most likely from a genetic mutation. The bird, like others migrating through Kiawah, fed on tiny clams along the beach and an inlet, Captain Sam's Spit. When Given and I walk the beach, we see neither the albino knot nor another unusual sighting reported here: when mullet are running, dolphin cooperatively "strand feed," first herding fish onto the beach, and then in synchrony launching themselves ashore, where they quickly feed before slipping back into the water. Knots with data loggers will corroborate Sanders's observations: two fattened up in the South Carolina spring, then flew directly to Canada.

Sanders is coming to think that knots may flock on Kiawah, feeding on clams until horseshoe crabs begin to spawn, at which time the birds will abruptly leave Kiawah. Knots have an uncanny ability to locate the best food. Researchers in Europe noticed that red knots "acted as if they had perfect knowledge about the relative quality of alternative foraging areas." Sanders has firsthand experience of this, watching knots disperse en masse from Kiawah after a spring full moon and then finding them gathering on beaches rich in eggs. On Harbor Island, where she and her team observed 2,000 knots, I watch spawning horseshoe crabs leave big, round nest impressions in the sand as they lay their eggs in the evening, and then the next day, with whimbrel flying overhead and a beach full of sandpipers, I see turnstones in the impressions, excavating sand and probing for eggs, while plump knots nearby pick eggs from the sand's surface.

Sanders has photographed them, scurrying along the beach with tiny eggs in their beaks, demonstrating that knots eat horseshoe crab eggs not only in Delaware and Massachusetts but in South Carolina as well, where other shorebirds—short-billed dowitchers, marbled godwits, and oyster-catchers—also share the feast. Observers on beaches in Alabama, Florida, Georgia, and New York find knots eating horseshoe crab eggs on these shores as well. John Tanacredi, a professor at Molloy College who monitors Long Island beaches from the tip of Brooklyn to the tip of Montauk, has seen knots, sanderlings, ruddy turnstones, semipalmated sandpiper, and willet—all eating horseshoe crab eggs.

Horseshoe crabs may matter to the lives of knots and other shorebirds not only in Delaware Bay, but along the entire coast. In South Carolina, knots have been eating horseshoe crab eggs for at least 100 years, even though ornithologists weren't describing it. In his study of shorebird food habits, Charles Sperry found that one knot, taken from Bird Island in Bulls Bay in the Cape Romain National Wildlife Refuge on June 7, 1915, "fed almost exclusively on spawn of the king crab, 110 whole eggs and fragments of many more having been found in its stomach." James Cooper, thinking of where to locate his new LAL manufacturing company, was attracted to Bulls Bay for the same reason as the knots.

Bird Island doesn't appear on the current refuge map, but horseshoe crabs still spawn in Cape Romain—on Bulls Island, where Wayne wrote about knots, and on White Banks Islands, Marsh Island, and other beaches as well. The broad, winding creeks, forested islands, and thick oyster banks in the sounds near Beaufort are a world apart from the shallow, open waters of Bulls Bay, dotted with a few tiny islands of shell and sand. One warm April evening, when the moon is nearing full and the tide is high, I'm taken into Bulls Bay to see horseshoe crabs. Sometimes high winds in Bulls Bay make it treacherous to take a boat through the shallow water. Tonight, it's calm.

We head out toward Marsh Island, known to some of the locals as Vessel Reef for the boats that can strand there. The island is packed with 13,000 shorebirds in the winter and many more during migration, including oystercatchers, pelicans, skimmers, and terns. We pull in and drop anchor. The island erupts in a cacophony of sound. As horseshoe crabs glide through the water to spawn at the island's edge, a fisherman collects and carries them to his skiff. They lie clicking on the bottom. Very early the next morning, they'll go to Charleston for bleeding and then be returned to the bay.

The sea reclaims its sand. Nearby Bird Shoal—perhaps this is the former Bird Island—has already disappeared beneath the peaking tide. Marsh Island, a smidgeon of sand barely one-tenth of a square mile, is shifting. Water laps at a peeling sign I'm told once stood on a small dune, now washed away. Bulls Island, where horseshoe crabs also spawn, is open to the public, but the refuge closes Marsh and White Banks Islands between February 15 and October 15 to protect nesting birds. The fisherman is here with a permit from the State of South Carolina and in violation of federal refuge regulations. The horizon blazes purple and pink as the sun sets. The moon rises. The tide is quickly ebbing; the crabs are leaving. We head back.

By the opening of the 2014 horseshoe crab season, South Carolina regulators, recognizing that the refuge had closed Marsh Island and responding to its concerns, amended its permit to prohibit fishermen from taking horseshoe crabs "in restricted areas designated by other entities of the State or Federal Government." Fishermen instead took horseshoe crabs from Bulls Island, an unrestricted area within the refuge, leaving Marsh Island birds undisturbed.

Another wildlife refuge went to court over horseshoe crabs. Cape Cod, Massachusetts's Monomoy National Wildlife Refuge, was, like Cape Romain, established as a sanctuary to protect migratory birds. Its mission, like Cape Romain's, is to protect and restore wildlife. National wildlife refuges allow recreational uses, but commercial users must apply for special permits, which can be denied if the use is incompatible with the refuge mission. As demand for horseshoe crabs began rising, a Massachusetts fisherman took horseshoe crabs from low-lying beaches and sheltered bays in both the Monomoy National Wildlife Refuge and the nearby Cape Cod National Seashore.

The Cape Cod National Seashore hadn't been aware that fishermen were illegally taking horseshoe crabs there. The Monomoy National Wildlife Refuge knew horseshoe crabs were being taken from its waters. In the 1990s, the refuge had permitted one fisherman to take horseshoe crabs for biomedical bleeding from an area closed to protect feeding shorebirds, having found, after a brief review, that impact was minimal. "Back then," remembers Bud Oliveira, retired refuge manager whose predecessor had issued the permits, "the companies were returning crabs, and no one understood how important their eggs were to birds." By 2000, shorebird populations and horseshoe crab populations were plummeting. "On Monomoy, horseshoe crab eggs are food for migrating birds. If you remove the crabs, you are removing food the birds are eating on their way back south." Oliveira declined to renew the permit. The Cape Cod National Seashore banned taking horseshoe crabs as well. When his permit was denied, the fisherman and the biomedical company he supplied with horseshoe crabs sued.

The judge ultimately upheld the bans. The following year the Monomoy National Wildlife Refuge conducted a lengthy analysis and determined that taking horseshoe crabs from the refuge was consistent neither with its mission as an "inviolate sanctuary . . . for migratory birds" nor with the national wildlife refuge mandate to conserve and protect wildlife and their habitats. The analysis pointed out that slowly maturing crabs, once heavily exploited, may not quickly recover; that declining shorebirds and their prey eat horseshoe crab eggs; and that there is little understanding of how collecting and bleeding horseshoe crabs affects their spawning success, fertility, or long-term survival.

These findings were made in 2002. More than a decade later, horseshoe crabs in Massachusetts are under stress: their numbers are declining; the proportion of male to female horseshoe crabs in Cape Cod's Pleasant Bay, where crabs are still taken for biomedical use, has increased; researchers are finding sublethal effects of bleeding that tire and disorient horseshoe crabs, and perhaps hinder successful spawning; and the red knot has been listed as threatened under the Endangered Species Act. Monomoy, reviewing its horseshoe crab prohibition 12 years later, proposes to continue the ban and, to further protect the prey of migrating ducks and shorebirds, to extend it to the commercial removal of mussels from the refuge.

We are but tenants on this planet, sharing the Earth with others who were here long before us and who, though less powerful, have tenancy as well. If there is any place where the well-being of horseshoe crabs, shorebirds, and other wildlife should be assured, it's in the nation's national parks, national wildlife refuges, and wilderness areas where, no matter how poorly we may understand or value these animals, they can live and thrive—for their sake, not for ours—and find sanctuary.

Shorebirds are by no means the only animals eating horseshoe crab eggs. During the few weeks they are spawning, they support a community of life, the energy from their tiny eggs felt throughout food webs at the edge of the sea. Silversides and killifish—small fish that spawn at the water's edge during spring high tides—gorge on horseshoe crab eggs and larvae, eating nothing else when horseshoe crabs are spawning. In turn, these fish nourish other animals. At the Monomoy National Wildlife Refuge, endangered roseate terns, down to 3,000 pairs, refuel on killifish and silversides before continuing on to South America. In the fall, sand shrimp in Massachusetts, fattened on horseshoe crab eggs, provide half the food consumed by migrating semipalmated sandpipers, who have miles to go before reaching their winter homes.

Horseshoe crab eggs feed juvenile striped bass, juvenile blue crabs, summer and winter flounder, perch, and eels. In July and August, off Long Island's western shores, horseshoe crab eggs and juveniles are the primary prey of juvenile striped bass. Eels in Rhode Island's Kickemuit River, a small river flowing into Narragansett Bay, eat fresh horseshoe crab eggs. An account from the nineteenth century reports that "the eels ... made a strange sight with their heads under the shell and their tails sticking out sideways." Take away horseshoe crabs, and an entire ecosystem weakens.

Dr. Al Segars, veterinarian for the South Carolina Department of Natural Resources and stewardship coordinator of the ACE Basin National Estuarine Research Reserve, lives on St. Helena Island, near Beaufort, on the edge of a marshy creek down a low road that floods in the spring high tides. He grew up on an inland farm. "If I wanted to play with kids," he says, "I had to walk a couple of miles. Instead, I walked barefoot to the pond." There is still nothing he likes better than being outside. Segars, Sanders, and I travel along the South Carolina coast near Beaufort into the ACE Basin, where the Ashepoo, Combahee, and Edisto rivers drain into the sea. We're looking for birds.

We take long walks to find them. We walk out onto the beach, where the distinguishing characteristics between western and semipalmated sandpipers are avidly discussed, the somewhat longer, slightly drooping bill of the western sandpiper a subtlety that still escapes me. On the way back, we walk by a large diamond-backed rattlesnake coiled beneath a porch. Unlike the sandpiper, the snake elicits little discussion. Segars takes us to ponds where wood storks roost high in the trees, on branches that seem too slight to hold birds four feet tall with five-foot wingspans. Closer to the ground, handsome black-bellied whistling ducks feed in another pond clogged with cattails and sedge. Segars tells me a pair of whooping cranes is wintering here in the ACE Basin. Eagles soar overhead. We walk by stands of cypress in black tannin-rich water, each tree surrounded by its mysterious brown knobby knees; huge, gnarled, live oaks, their branches draped in moss and covered with resurrection fern; flinty-eyed heron hidden among reeds and dozing alligators. I am lulled into thinking I'm walking in a lush, ancient, and untouched wilderness.

That's not quite the case. Cypress is thinner here than it once was. Forest upon forest of longleaf pine were tapped for pitch and turpentine and felled for timber, sending the red-cockaded woodpecker toward extinction. The pines are being replanted and the midstory burned to control growth. Sanders's husband drills nesting cavities for red-cockaded woodpeckers into the trunks of older trees. The dense wood stork rookeries are relatively recent; 11 nests recorded in 1981, and now 1,500–2,000 each year. Whistling ducks and whooping cranes are newcomers as well, the first crane arriving after being blown off course en route between Florida and Wisconsin. White pelicans, long seen in South Carolina in small numbers, are now so numerous a distant pond where they alight seems covered in falling snow. All these birds live in managed wetlands whose water levels people carefully control.

The succession of events that sculpted today's landscape along South Carolina's estuaries began more than 300 years ago when slaves working with axes and shovels, up to their waists in water and their knees in mud, cleared bottomland forests and malaria-infested swamp to grow rice. In impoundments diked along rivers touched by the sea, they set thick rectangular culverts 20 to 30 feet long, made of cypress, pegged with risers, and gated at both ends—classic Lowcountry rice "trunks." Raising and lowering the gates, and fine-tuning the water flow with risers, trunk minders flooded and drained the fields on incoming and outgoing tides, tasting the incoming water for a surfeit of salt. The nutty-flavored Carolina Gold rice created great wealth. By 1770, coastal South Carolina held one of the greatest, if not the greatest, concentrations of wealth in North America.

By 1900, the South Carolina Lowcountry had become one of the country's poorest regions. The Civil War, the subsequent loss of skilled slave labor, a market flooded with competitively priced rice from Texas, Arkansas, and Asia, and hurricanes that battered dikes and flooded fields with saltwater devastated the area. Some of its plantations were sold and developed into resorts. Hilton Head is one. Others remained intact, purchased by new elites, people like Max Fleischman of the Fleischman Yeast Company and Gaylord Donnelley, owner of the world's largest commercial printing company, based in Chicago. The new owners used the plantations as private hunting preserves, repairing rice impoundments and trunks to attract ducks and geese. "Instead of growing rice," Segars says, "they grow wildlife."

Today, collaboration among the South Carolina Department of Natural Resources, the U.S. Fish and Wildlife Service, Ducks Unlimited, the Nature Conservancy, other conservation organizations, and private landowners keep wetlands in these old rice fields intact. Protected lands include, but are hardly limited to, Atlanta media executive and philanthropist Ted Turner's plantations, under private conservation easements; the Dupont lands, managed by the private, nonprofit Nemours Wildlife Foundation; the former Santee Gun Club (first saved from insolvency in 1900 by Boston department store magnate E. B. Jordan, who named one of its duck ponds Jordan Marsh); and the lands of Tom Yawkey, former owner of the Boston Red Sox.

Hunters help pay for this protection. Throughout the United States, $7 billion in excise taxes on guns and ammunition levied under the Pittman-Robertson Wildlife Restoration Act have purchased, leased, or acquired easements on 5 million acres to protect wildlife, enhanced habitat on another 38.6 million acres, and provided wildlife management assistance to over 9 million landowners. Over $1 billion of revenue from Duck Stamps—required for duck and goose hunters—has enabled the U.S. Fish and Wildlife Service to add over 5 million acres of wetlands to the nation's national wildlife refuges. In South Carolina, hundreds of thousands of coastal acres supporting hundreds of thousands of birds are protected—approximately 28 percent of the coast within reach of the tide, according to calculations made by Michael Slattery of the South Carolina Sea Grant Consortium and Coastal Carolina University.

South Carolina has 70,000 acres of old rice fields. Managed wetlands in the Tom Yawkey preserve have long attracted thousands of shorebirds, including as many as 32,000 semipalmated sandpipers in the spring. Shorebirds there favor rice fields over tidal mudflats 16 to one. Nathan Dias, who conducts detailed weekly shorebird surveys in the preserve during spring, fall, and winter, has seen 32 species of shorebirds in the former rice fields, including semipalmated, western, pectoral, and least sandpipers, short-billed dowitchers, killdeer, willet, whimbrel, red knot, and ruddy turnstones: all shorebirds either endangered or whose numbers are dropping. In Yawkey, Dias says, drawdowns are staggered to drain at least one or two impoundments every month, "purely for the benefit of shorebirds."

Shorebirds are benefiting. In Yawkey's former rice fields Dias has recorded knots and ruddy turnstones in increasing numbers, and a few firsts: a pair of roosting American oystercatchers taking refuge from a storm and its accompanying high tide, and beach-nesting Wilson's plovers, nesting and fledging their young in the drained rice fields. On May 3, 2014, he counted 50,000 shorebirds in the Yawkey ponds, and a few weeks later, 56 red knots, a record count for the impoundments. "Managed impounded wetlands," one study finds, "are credited with providing important habitat for shorebirds as natural habitat declines." Ernie Wiggers from Nemours hosted a workshop for Lowcountry plantation managers on how they might use impoundments to attract shorebirds. (Barbecue was served.) With shorebird populations dropping, impoundments, "the table that's set all day," as Segars describes them, may help relieve the stress.

Duck Stamps and excise taxes on guns, bows, and ammunition conserve game birds by guaranteeing permanent, dedicated funding to protect their homes. Hunters are powerful advocates for conservation. One analysis described their effectiveness this way: "Historically, about 90 percent of state fish and wildlife agency budgets have been derived from fees on hunters and anglers to manage less than 10 percent of the species under their jurisdiction . . . a long-standing funding imbalance."

Segars believes that just as hunters contribute to the protection of game species, birders should help protect nongame species. He recommends that people buy Duck Stamps. "If you use these places," he says, "you should help support them." I've heard this opinion expressed repeatedly, not only in South Carolina but also in Delaware and New Jersey. Niles, never one to mince words, is blunt. "Hunters pay to take their shot," he says. "Birders and wildlife photographers, 'shooting' through a lens, should share the cost of wildlife conservation as well."

Many are willing to do just that, but proposed excise taxes on binoculars and spotting scopes, field guides, bird feeders, bird feed, and recreational outdoor hiking and camping equipment—intended to right the funding imbalance and supported by hunting and conservation organizations, optics manufacturers, and sports fishing organizations and retailers—failed in Congress, opposed by the Outdoor Recreation Coalition of America, a recreational company trade group, and the large recreational outfitter REI.

Shorebirds, and horseshoe crabs, are being squeezed out of their seaside homes. Walking the beaches of Cape Romain, where low-lying islands are washed by tides, it's hard not to feel their ephemeral nature. Barrier islands are built on shifting sands. Currents and waves raise dunes, and hurricanes wear them away. Sand disappears from one shore and piles up along another. In Cape Romain, jetties, dams, and river diversions to the north withhold sand, starving the refuge's barrier islands. Year by year the losses exceed the gains. Cape Romain's early settlers grazed cattle on an island beach known as the Cow Pens. It's now worn to a thin ribbon of sand shrinking 20 feet each year. Nearby Sandy Point, which Wyndham tells me once hosted the highest dunes in the area, is no longer sandy, no longer a point. When it disappeared, an important nesting site for skimmers and terns disappeared as well, forcing the birds onto the beaches that remain. The sea is rising in South Carolina, as it is along much of the eastern seaboard. In Cape Romain, rising water exacerbates a situation already made difficult by the northern jetties and dams. With a three-foot rise in sea level, Cape Romain will lose most of its wide expanse of marsh and beach to open water. Already, as the beaches wear away, refuge staff move inundated loggerhead nests to high ground. Shorebirds and horseshoe crabs may lose their beaches as well.

More than half the South Carolina beaches where red knots feed are highly vulnerable to rising water. At the same time, 40 percent of the state's dry land and freshwater wetlands within three and one-third feet of sea level is undeveloped. If the water doesn't rise too quickly, marsh and beach may have room to retreat. Wyndham sits with South Carolina's old maps and histories, piecing together the ebb and flow of sand. His house and the porch where he works are elevated—enough to park a car beneath and perhaps to weather a storm surge from the next big hurricane. He faces the fleeting nature of the home and land he loves with equanimity. Standing with a friend on the man-made dikes of an old rice field with its carefully calibrated water levels, he watches 17 ibis circling the pines, glowing in a flash of afternoon sunlight. He is moved by the aching beauty of marsh and beach and wetland continuously redesigned by humans, a glorious trace of what must have been, and now passes, perhaps in our time. I see him, if not in this life, then perhaps in another, sitting on his porch, the marsh in his yard, the beach and sound at the end of his driveway, and the long-billed curlew still standing, still watching, still home.

In the lowlands of Tidewater Virginia, the beaches where Barry Truitt and the Nature Conservancy work, and where knots feed in the spring, are also on the move. On Hog Island, where Truitt saw knots from South America and Canada, the town of Broadwater once nestled in a pine forest at the island's center. During its heyday, 300 people lived there, enough to support 50 houses, a school, a church, and a handsome hunting club and shooting lodges built by friends of President Grover Cleveland. As the island rotated, Broadwater slipped into the sea, and now rests hundreds of yards from shore. Today the island is uninhabited and can move freely.

Beaches along Delaware Bay, a critically important spring stopover for migrating shorebirds and home to the greatest concentration of spawning horseshoe crabs, have been washing into the sea for at least 150 years. Maurice Beesley's history of Cape May describes 300 acres of Egg Island, where Alexander Wilson saw so many horseshoe crabs, washing into the water. William Kitchell, superintendent of the New Jersey State Geological Survey, writing on the state's geology in 1856, collected accounts of erosion at the edge of the sea: an island in the meadows of Goshen harboring trees in 1786, but 70 years later covered in four feet of muck and marsh and inundated by high tides; cedar swamps 11 and 17 feet below what are now salt marshes. The encroaching salt marsh is so common, he writes, that "observations on the dying out of timber can be made by anyone who is interested to do so."

He also describes rapidly eroding beaches: sandbanks 15 to 30 feet high once covered with trees, leveled; a mile of beach in front of the Cape May boardinghouses, where the militia practiced during the Revolution, all gone, and the boardinghouses twice moved inland; a town on the bay side and behind it, inland, a cemetery, where over the years, Aaron Leaming watched the houses wash away, sand blow into the cemetery, and the grave of his grandfather, one of Cape May's first settlers, wash into the sea. Today, the site of Cape May's Portsmouth Town, settled by whalers in 1640, rests half a mile out in the bay. It's not all loss—Kitchell describes places where the sea has receded, where he finds shells buried beneath the soil—but overall, he concludes that Cape May is sinking.

For Willets Corson Camp and his sister, thinking about changes on Delaware Bay since their childhood, when their father and grandfather ran the horseshoe crab fertilizer factory, it's the erosion that stands out. "The tide took one house out," says Frances Camp Hansen, "and it lapped right up to another. Our house is gone, but if it were still standing, it'd be out in the middle of the bay." Her brother agrees. Houses continue to slip into the sea; the bay has more than one ghost town.

One early spring day, when the trees are red with buds and ospreys are flying back to nest, I walk out to Delaware Bay's Thompson's Beach on a mile-long road cutting through a marsh where, in 1685, Thomas Budd fed his cattle and cut salt hay. The narrow beach is filled with rubble: cinder blocks covered with algae; chunks of concrete strewn along the shore; torn-up paving from an old driveway; a broken chimney; pilings from old wharves; bulkheads that ultimately failed to fend off water. Buried in the sand I find a dirty but intact glass electrical insulator made from New Jersey sand at the former Whitall Tatum glass factory in nearby Millville. The beach has eroded, exposing the peat below. Sand has moved inland, lining a creek behind the old house foundations. I walk by a few horseshoe crabs that spawned, or tried to spawn, among the cinder blocks, and died. In November 1950 a blizzard destroyed all but a few of the 88 homes on Thompson's Beach.

Reeds Beach, one of New Jersey's best horseshoe crab spawning beaches, is thinning. Two rows of cottages, separated by a road covered with sand, squeeze between marsh and shore. The dunes, pushing to the edge of the road, are fenced, but sand nonetheless spills across the road and into the marsh. The beach is migrating, but it's blocked by houses and bulkheads designed to keep the water at bay. The beach washes away, exposing peat below the sand. Horseshoe crabs skirt the peat and spawn beneath houses raised on pilings. There's not a lot of room here. Hurricane Sandy took another swipe at Reeds Beach, leaving some houses unscathed, but ripping porches and entire rooms from others, turning them to splinters in the sand.

Hurricane Sandy destroyed 70 percent of New Jersey's prime horseshoe crab spawning beach. Niles and Dey had already seen a 2003 storm ruin the spawning season and leave birds hungry. There were too few birds to let this happen again. Niles is a tenacious man. In five months, a coalition of scientists, government agencies, and conservation groups, including the New Jersey Division of Fish & Wildlife, the American Littoral Society, the Conserve Wildlife Foundation of New Jersey, and the Wetlands Institute, raised $1.4 million, applied for and received the necessary permits, hired contractors, identified suitable sources of sand, removed 800 tons of debris and layered 40,000 tons of sand on five beaches, covering the peat shelves and restoring more than a mile of beach. Both birds and crabs favored the restored beaches over those still damaged. In 2014, the groups received another $1.65 million to remove more brick, pilings, and asphalt, and to bring in another 45,500 tons of sand. The work will continue in 2015 with another grant of $4.7 million, given by the Department of the Interior to help improve the community's resilience to future storms.

Across the bay in Delaware, Richard Weber, counting horseshoe crab eggs on Delaware's beaches, had taken me to Port Mahon, once a great horseshoe crab spawning beach, and now, he says, "Delaware's ugliest beach. It used to be wide and uniform, with deep sand. We're losing it. Less than

20 percent of the spawning area is left." Over the years, Delaware has rebuilt its eroding beaches, adding over 3 million cubic yards of sand in grain sizes that appeal to horseshoe crabs. "As long as we take care of beaches, crabs will take care of birds," Weber tells me. When shorebirds began losing their seaside homes, the man-made jetty in Mispillion Harbor may have helped save the red knot from extinction. The jetty shelters the beach from winds and waves, creating a gentle environment conducive to horseshoe crab spawning and producing the highest concentration of horseshoe crab eggs in Delaware Bay.

In Delaware, planners anticipate that a three-foot rise in sea level may inundate more than 80 percent of all the state's wetland impoundments and national wildlife refuges, affecting more than 97 percent of its tidal wetlands. A four-foot sea-level rise may inundate almost all the impound-ments and national wildlife refuges, and alter almost all tidal wetlands. Already, flooding on Delaware Bay beaches during May high tides may have impeded horseshoe crab spawning.

I love walking out on Delaware's Slaughter Beach on dark nights when horseshoe crabs are coming in with the tide. I stand in the quiet, listening to waves lap against the shore, watching female horseshoe crabs burrow into the sand and then emerge as the tide falls. The line of crabs extends along the beach as far as I can see. One windy night, the beach is empty when I arrive. The wind is churning the surf, the waves are too high, and the crabs are staying offshore. Another night is calm and still, but every-thing—beach, marsh, wrack line, even the peat ledges—is flooded. The water is almost at my car. The horseshoe crabs stay away on this night, too. Like New Jersey, Delaware is hoping to increase the resilience of its shore-line. $6.5 million in grants for 2015 will go toward rebuilding marsh where shorebirds could roost, moving wildlife impoundments inland, and repairing the Mispillion Harbor jetty and spawning beaches.

The repairs on both sides of the bay are buying time—for horseshoe crabs, for shorebirds, and for people. How long they will last is an open question. Storms and currents will eventually carry away the new sand, typically in two to six years. The years ahead promise great changes along Delaware Bay. In 1953, Thurlow C. Nelson, then chairman of the New Jersey Water Policy and Supply Council, spoke of stresses on New Jersey's water supply, which he attributed to newfangled washing machines and dishwashers, pollution, and development, and which he believed were worsened by a recent nine-inch rise in sea level caused by melting glaciers. As the Earth continues to warm, scientists project the sea level may rise three to six feet on New Jersey's bay side. By 2050, severe 100-year storms will batter the shore every 10 to 20 years. In 1950, the storm surge brought by Hurricane Sandy would have been a freak event, occurring once in 1,500 years. By 2012, the odds had increased to once every 500 to 700 years. By 2050, Hurricane Sandy storm surges may flood New Jersey shores every couple of years.

The old histories of Cape May describe again and again forests and freshwater wetlands yielding to salt marsh and sea. Horseshoe crabs have followed rising and falling coastlines for millions of years. When the last Ice Age lowered sea level by 400 feet, they still spawned, most likely on a shore now out at sea. When the glaciers retreated and the sea level rose, flooding the mouth of the Delaware River, they followed the rising water, laying their eggs on retreating beaches. Through it all, they survived. How they will fare in the years and decades ahead with so much development along the shore is hard to say. Global warming may be presenting knots with their first big bottleneck since the last Ice Age. Niles began his professional career acquiring land to protect shorebirds. As the water rises and the beaches Dey and he have come to know so well over so many years give way again and again to Hurricane Sandy–type storm surges, he may find himself coming full circle, once again working to secure horseshoe crabs and shorebirds new homes, on a new shore, farther inland.

Georgia's Altamaha River flows untrammeled 137 miles to the sea, carrying sand and sediment from cypress and tupelo swamps and old-growth longleaf pine forest, dropping it on islands and sandbars as the river slows and empties into the ocean. I'd come to one of these islands, Little St. Simons, hoping to see horseshoe crabs, wondering if I'd see knots. On September 18, 1996, biologist Brad Winn put the sand islands at the mouth of the Altamaha River on the fall migration route for red knots. It wouldn't be the first recorded large sighting of knots in Georgia. Brian Harrington, scouring the East Coast for knots, had found a 1971 account from Herman Coolidge, a birder and bankruptcy judge from Savannah. Coolidge, riding along the beach on nearby Wassaw Island, came across what he concludes were "at least 12,000 knots." Shorebird surveyors went back in the following years but never saw them again. Even earlier, in 1890, self-taught taxidermist and collector Wills W. Worthington shot a few knots at the mouth of the Altamaha.

Winn had taken conservation groups and Georgia sea island managers to a cluster of islands there—Wolf Island, Egg Island, and Little Egg Island Bar—to show their importance for seabirds and shorebirds: nesting American oystercatchers and Wilson's plovers; wintering or migrating whimbrels; piping, black-bellied, and semipalmated plovers; dunlin. Before they even got out of the boat, Winn saw a large flock, flying low in tight formation over the bar. "For the next hour," he remembers, "they put on quite an aerial display. There was clearly a peregrine nearby, although I never saw it. The birds were agitated, landing, then quickly taking off, flying so low we could hear the sound of their wings and feel the river of air over our heads." Winn had unexpectedly come upon 5,000 red knots.

They were eating dwarf surf clams. When clams are abundant, so are knots. Biologist Timothy Keyes from the Georgia Department of Natural Resources tells me that in good years, when tiny clams are abundant, red knots snap them up at a rate of one every two and a half seconds. In years of plenty, Keyes has been knee-deep in windrows of clams 100 yards long. He tells me that in years of drought, when less freshwater comes down the Altamaha, salinities are higher, clams are fewer, and the knots find them more slowly—one every 30 to 40 seconds. In the fall of 2011, Keyes and his colleagues estimated that 20,000 red knots heading south came to refuel on the islands at the mouth of the Altamaha.

I'd come to Little St. Simons in the spring, though, and wasn't sure what I'd find, although shells of some of the largest horseshoe crabs I have seen come from this quiet island, whose owners include Wendy and Henry Paulson. As a child, Wendy Paulson would accompany her father on casual birding walks, and then as an adult, she came to know birds by their songs, a method of learning she describes as "Berlitz birding." A self-taught birder aided by the generosity of friends, she saw her first knot on Assateague Island in its gray winter plumage and felt she'd "never, ever recognize this bird that was so indistinct"—until she saw it again in russet breeding plumage. Returning the generosity extended to her, wherever she has lived she's led bird walks, encouraging others confused by difficult-to-distinguish shorebirds and, coming to love the knots she'd seen, supporting Rare's work to dramatically increase people's awareness and appreciation of knots in Argentina—in Río Grande, Río Gallegos, and San Antonio Oeste.

There is lots of wildlife to see on Little St. Simons: armadillos, pileated woodpeckers, and chuck will's widows in the woods; painted buntings at the feeders; a roseate spoonbill and egret rookery, thick with nests and baby birds and protected from raccoons by large and numerous alligators; a field of ibis (one day I counted 400); a pond of a dozen stilts with their astonishingly long bubblegum-pink legs; wood storks, tricolor and little green herons, and rails. Oystercatchers, willets, and yellowlegs, and Royal, Caspian, gull-billed, sandwich, and least terns all roost or feed on long stretches of undisturbed beach. I accompanied graduate student Abby Sterling on her search for Wilson's plover nests, following their tracks into the dunes.

Horseshoe crabs had just begun spawning. Knots, which hadn't been on the island in large numbers, had just arrived. Early one morning, we see 1,000 feeding on a thin sandbar just off the beach, along with a few marbled godwits. The sea is rising here too, but with little to hold back the sand, the beaches are free to shift with tides and currents, and to move with the ebb and flow of water. Little St. Simons, Wolf Island, Egg Island, and Little Egg Island Bar are nourished by sand from the Altamaha. The beach we walk has grown 200 feet in just five years. The birds are at peace. Here, at least, they can follow the beaches as they migrate.

While I was following knots, South Carolina and Georgia emerged as important stopovers. Knots are shorebirds, but their route north doesn't always follow the shore. I make a detour before going to the Arctic, to be with researchers as they find another route, inland, a ghost route, no longer well traveled but still in use.

Nine

GHOST TRAIL

The Laguna Madre and the Central Flyway

t's December in Corpus Christi, Texas. A bank of security cameras lines both sides of the road at the gatehouse to the Padre Island National Seashore. A sign warns that driving conditions on the barrier island beach are poor. On the way in, David Newstead and I stop at a pond. It's filling with redheads, pretty ducks with dark red heads and blue bills, so many that I can hear the flit-flit of their wings as they land. I've never seen hundreds of redheads, and there'll be more to come.

The pavement ends just past the visitor center. "Red knots were here in the fall," says Newstead, who is driving. He majored in English in college, has a master's degree in marine biology, and works for the Coastal Bend Bays and Estuaries Program. He specialized in fish and fish larvae, but now he's hooked on birds. He began his work with colonial nesting birds—the reddish egrets, roseate spoonbills, and black skimmers—that breed on small islands throughout the Laguna Madre, behind the barrier island. "They are easier to deal with. They're local. You know their histories. If there's a problem you stand a better chance to figure out what and where it is. Migrating birds are different, harder." Newstead says he hopes that soon he'll "crack the case" of where the red knots he sees on Texas beaches in the fall spend the winter.

I'm here to see how he's doing it. I nearly didn't come to Texas. It wasn't considered a key stop on the knot migration route. Back in 1992, when the population was much higher, even the all-knowing Harrington hadn't seen that many birds there—1,400 in November, but winter flocks of only 100–300 birds. Still, I'd heard about Newstead's work and was curious about why he was looking for knots and how he was trying to find them.

Padre Island, the world's longest barrier island, extends down to the mouth of the Rio Grande at the Mexico border. The first 60 miles, down to the Mansfield Pass, one of several manmade cuts in the island, belong to the National Park Service. Today we're going partway. Northeast winds and a recent storm have pushed the water right up to the dunes. Until the tide smoothes out the beach, the going may be a little rough, requiring four-wheel drive to get us through the soft sand.

Newstead counts snowy and piping plover as he drives. More than half the piping plovers that breed in the Great Plains of Canada and the United States winter along the western Gulf of Mexico. We pass stretches of beach covered with tiny clams. The shells glint blue, gold, white, and pink in the sun. "There's two species here," Newstead says. "I've seen a recipe for some kind of stew. It'd be a lot of work." The clams are smaller than my smallest fingernail. At the 10-mile mark, the sand grows coarse—Little Shell Beach. A dolphin carcass lies washed up in the sand. At the 25-mile mark, the sand grows coarser still—Big Shell Beach. In the summer, sea turtles nest here. Their history—of abundance, decimation, and restoration—is imaginable for knots.

Sea turtles once filled Texas bays and lagoons: in 1890, 1 million pounds of sea turtles were caught in the Gulf of Mexico. Texas took more than half. Canneries, often associated with large meat-packing plants servicing nearby cattle ranches, shipped turtle meat and turtle soup north. Eggs were prized aphrodisiacs. Turtles quickly disappeared, although in the 1940s an engineer, Andrés Herrera, flying back and forth along the Mexican coast searching for Kemp's ridley sea turtles, found, on the twenty-sixth day of his search, 40,000 of the now-endangered turtles. They were coming ashore, en masse, to breed on the beaches at Rancho Nuevo, Tamaulipas, Mexico. For an entire mile, he reported, the beach was saturated with eggs, covered with turtles. There are old reports of Kemp's ridleys nesting in the sands of Padre Island's Little Shell and Big Shell, remnants of mass nesting—arribadas—that may have once taken place there as well.

By the 1970s, prospects for Kemp's ridley sea turtles were grim. Today, decades of dedication and hard work in both Mexico and the United States are yielding results. The beaches of Rancho Nuevo are now a protected sanctuary, and shrimpers are banned from nearby waters during nesting season. The National Park Service, assisted by hundreds of volunteers, protects nests and incubates Kemp's ridley eggs on Padre Island. Only a few hundred turtles nested in Rancho Nuevo and none were known to still nest on Padre Island in the 1980s. By 2012, between 7,000 and 8,000 Kemp's ridleys were crawling onto Gulf of Mexico beaches to lay their eggs, with between 70 and 85 on Padre Island. With these protections, the Kemp's ridley may, in my lifetime, be pulled from the brink of extinction and once again thrive. What's possible for sea turtles is possible for shorebirds.

Newstead leads the way up a dune, walking through sea oats whose wide-spreading roots penetrate deeply into the sand, anchoring the 40-foot dune. Below us, between Padre Island and the mainland, are the tidal flats and shallow waters of the Laguna Madre, the Mother Lagoon, a long, narrow body of water, which together with its Mexican counterpart south of the Rio Grande—the Laguna Madre de Tamaulipas—is 230 miles long, the largest, and one of the saltiest, lagoon systems in the world. We're facing a wide, green mat of algae where knots gather. At its center is Nine Mile Hole, where redfish can be trapped on an incoming tide and Newstead sometimes fishes. "In the summer the water is low. The sun bakes the flat lifeless. The mat's as hard as rock and there's no food, so the birds are on the beach. Come October, the flats are inundated. When it's ankle deep, there'll be shorebirds and a few hundred heron." Today we see western sandpiper and dunlin. No knots. A little farther back on the beach, we'd hiked over to another mat where 1,000 knots had been seen one spring. We didn't see any there either.

Most American knots using the Atlantic flyway winter in South America. A smaller segment winters in the southeastern United States, primarily along the Gulf Coast of Florida. I'd gone to see them there, at Fort de Soto State Park on Tampa Bay in St. Petersburg, on a string of beautiful sandy beaches. Reddish egrets, great blue herons, and little blue herons fished in mangroves behind the beach. Ibis walked in the grass, around picnic tables not yet filled with families. Skimmers, terns, and oystercatchers roosted on a small sandbar across a shallow inlet, along with about 20 marbled godwits, a shorebird I hadn't seen in particularly large numbers. Walking the beach, we finally found knots—80 or 100—in a small fenced-off area where they wouldn't be disturbed. I hoped there were clams for them to eat inside the fence. We stood with a long-billed curlew and watched them. The number of knots in Florida, 6,000 in the 1980s, has dropped to between 1,000 and 1,500.

The birds frequent Padre Island and Mustang Island to the north in both spring and fall—the National Park Service has reported 1,600 in a day, and a friend of Newstead's, Captain Billy Sandifer, has counted as many as 3,000—but they had seen very few flagged birds from other places. Newstead had seen one on Canada's Mingan Islands. With perhaps one-quarter of the birds flagged, this is odd. "We need to know what's going on," Newstead says. He thinks knots might winter here, a different group from those flying along the Atlantic. So far he has little evidence.

The beach is lined with dead fish from an awful red tide: ladyfish (skipjack), redfish, pompano, whiting, mullet, catfish, stargazer. Newstead points to what's left of a catfish. "Look at those spines. They'll go right through a tire." He'd bought a new truck just before dead fish began washing up on the beach in September, and it'd been impossible to avoid driving over them. As he did the fish exploded. Their guts sprayed up under the hood and cooked. "The stench was horrible." More dead fish would wash in with each tide. It got to the point where without getting out of his truck, Newstead knew how long they'd been there. "Day one, they crunch. Day two, they pop. Day three, the wheels slide right over. It sounds morbid, but it's the only way to get through." A large redfish, recently washed in, lies on the beach, its eye sockets empty, picked by gulls.

Sandifer, who lives in a mobile home in nearby Flour Bluff, knows Padre Island well. When Newstead and I visit him, he is getting ready to make his annual Christmas tamales. His small, shady yard is filled with plantings to attract birds. Spring migrants, exhausted from a rough trip across the Gulf of Mexico, drop into his yard. He's been known to exercise control over problematic invaders with a pellet gun. Sandifer fought in Vietnam. He spent seven years, five months, and 18 days in the service and returned ill. Padre Island took the edge off his anger, restored his spirit. For many years he held the only concession from the Park Service to take people surfcasting on the beach. "Every year for 22 years I've been on this beach, day after day. It saved me."

Fishing is how he got into birding. "Some Yankee came down here, a fine person, an exceptional birder, too damn cheap to hire a guide. We made a deal: he'd teach me about birds, I'd take him fishing. He really did me. I'd say, 'It's a semipalmated [plover],' and he'd say, 'Why isn't it a snowy?' All the birding newcomers want to do is find something unusual. That's not how it works. Instead you need to know all the birds that are supposed to be there, and then the rarity stands out." Sandifer, according to Newstead, knows Padre Island better than anyone else, knows his birds so well he finds both rarities and birds that turn out to be more frequent visitors, but that he is the first to see. He's been watching knots for years.

"I used to see them in the fall, flocks of 1,200 to 1,500 birds, day after day, down near the 15- or 20-mile marker. They were a sign of the season. The most I ever saw was about eight or 10 years ago, in the spring. The wind was about 45 miles per hour on the beach. I'd taken people fishing at the Land Cut, where it was protected." Over the years, 20 miles of sand and mud—the Saltillo Flats—built up, dividing the Laguna Madre. The Land Cut, dredged for the Intracoastal Waterway, reconnected the bays. "I saw knots on a sandbar emerging from the water. Not only were they fat, there were thousands of them. They're less frequent now. Now it's a few here, a few there. Things have changed. I used to go 16 days on the beach without seeing another human being or a car. Now it's 16 minutes."

The massive fish kill isn't helping either. "This is the worst red tide I've ever seen. It's the longest duration, and the most widespread, extending from High Island and the Bolivar Flats near Galveston, down to the Rio Grande." By the time it dissipates (after five months), it will have been one of the worst recorded in Texas history, killing 4.5 million fish and temporarily closing all the state's oyster beds, costing the industry $7 million.

The red tide is borne in on a tiny alga, *Karenia brevis,* which, when it blooms in profusion, turns the water red or tea colored. Normally the algae live offshore, but there's little freshwater emptying into the ocean here: months of drought have increased the sea's salinity. Wind and currents concentrate the algae closer to shore, where they thrive in the saltier water. *Karenia brevis*'s potent toxin can paralyze the central nervous systems of fish; ingesting it in lethal doses, they gasp for air and suffocate. Red tide kills crabs, shrimp, dolphins, even coyotes that eat infected fish. Toxins can accumulate in shellfish feeding in red tides. While they are unharmed, humans, birds, and sea turtles eating tainted shellfish can become seriously ill.

Red tides have bloomed in the Gulf of Mexico for hundreds of years. One of the oldest accounts is from the Spanish conquistador Álvar Núñez Cabeza de Vaca. Defeated in his attempt to claim the coast of the Gulf of Mexico for Spain, he ordered his men to slaughter their horses for food. They fled from Florida by raft. A hurricane blew them to south Texas, where they lived for several years, Cabeza de Vaca noting in 1534 that Indians marked the changing seasons by "the times when the fruit comes to mature and when the fish die." In 1648, Fray Diego Lopez de Collogudo, a Franciscan monk living in the Yucatan, wrote of a "foul odor" blowing in from the sea to Merida. The source: "a mountain of dead fish" heaped on the shore. Red tides in the Gulf of Mexico are as old as the area's recorded history. They're increasing in severity, duration, and geographic extent. Sandifer is not surprised. "You can't mess with the top of the food chain, and you can't mess with the bottom either."

What is perhaps surprising is the complexity and reach of the disturbance. Increasing aridity and desertification in Africa reverberate across the Atlantic in the Gulf of Mexico as, each summer, churning Saharan sandstorms blanket the gulf with dust, seeding the sea with iron. At the bottom of marine food webs, floating colonies of the tiny microbe *Trichodesmium* soak up the iron and bloom. Their growth enriches the water with nitrogen, which in turn feeds the red tide. At the top of the food web, humans remove the larger fish—red snapper, grouper, and mackerel—and then remove their prey of sardines, herring, anchovy, menhaden, and pink shrimp. Now with fewer predators to graze them, the floating plants of the sea, some of them toxic, grow more profusely.

Karenia brevis is often in gulf water at concentrations too low to harm. When their numbers rise to as many as 25 cells in a teaspoonful of water, shellfish beds close. The patches of red tide that I could see from the beach could have millions of cells in a single teaspoon. Humans, even if they aren't eating shellfish, feel the effects. In rough surf, the algal cells break up, releasing toxin into the water. Aerosolized in the salt spray, it's an eye and throat irritant. "It was really bad," Newstead recalls. "Driving, I had to keep the truck windows rolled up. Out on the beach, we had to wear masks. We were all hacking. Knots were lying in tire ruts in the sand. They looked as if they were about to die." One did. He took it to Paul Zimba, harmful algal bloom biologist, who found toxin levels in its liver at 16 times the lethal dose.

Red tide is by no means unique to the Texas coast. It occurs regularly, often for months at a time, along the Gulf Coast of Florida where knots winter. In 2013, the National Weather Service added an alert, warning when red tides in Tampa Bay threaten human health. In the spring of 2007, 300 knots died at Uruguay's Playa La Coronilla, a beach near the Brazil border. Another 1,000 dead shorebirds, of unknown identity, were found nearby on the same day. Given the time of their arrival, the knots were most likely en route from Tierra del Fuego. Local birders think a red tide may have killed them, but their cause of death was never determined.

The other knots suffering from red tide in the laguna went to rehab, to ARK—Animal Rehabilitation Keep—run by another old friend of Newstead's, Tony Amos, in Port Aransas. I meet him at dawn on Mustang Island, a barrier beach 18 miles long, north of Padre Island. It is dark when I set out. As I inch the small Toyota I've borrowed from a friend out onto the highway, pickups whiz by. The speed limit seems to be a minimum here. I decide to avail myself of this opportunity to drive at speeds frowned upon in Massachusetts. With the extra time gained, I pull over at a pond packed with redheads. More than 90 percent of the entire population winters in the lagunas and bays along the Gulf of Mexico. The sky begins to lighten, but I linger. These birds are rare where I live: I love seeing so many, and so close.

The beach access road, like others I'd drive on this trip, has no parking lot. I find campers and trucks out near the dunes, and park there. When Amos arrives, we head out in his truck to scan a seven-mile stretch of this gulf beach. He counts and records everything he sees. Amos is in his seventies. He's been doing this survey every other day for 35 years.

Right away we see two herring gulls, three laughing gulls, five ring-billed gulls, two willets, two ruddy turnstones, one western sandpiper, four black-bellied plovers, two walking people, four people camped in a car, and three empty carboys. According to the labels, they once carried caustic alkali liquid. "They came off a boat," Amos says. Each is numbered and has a date. He'll look them up later. At the first half mile, he gets out and walks the width of the beach, recording how many paces to the high tide line, to the dunes, to the mile marker. "I've logged 1,700 miles walking the width of this beach."

We resume the ride. He makes estimates of washed-up turtle grass from the bay and Portuguese man-of-wars from the sea. He calls out the count: one willet, two dunlin, a black-bellied plover, seven ringbills, 10 herring gulls, one beverage, four great-tailed grackles—they come to eat the insects in his truck—a great blue heron, two ruddy turnstones, a sanderling, 10 piping plover, three milk jugs, two western sandpipers, one laughing gull, and an American oystercatcher. He counts single-drink beverage containers, green bleach bottles from shrimping boats, and egg cartons. He counts dogs and mangrove seeds. "I don't like dogs on the beach, especially off the leash. They disturb the peace and flush birds." He puts on gloves to look at a dead red-tailed hawk. "Things do die on a regular basis. The chest cavity is stripped. Some birds I've only seen dead."

He's looking for piping plovers. "They're holding their own here." He sees one, takes out his camera, zooms in to photograph the number on the flag. "I know this bird. I first saw it in July 2010. It's from the Canadian Great Plains, and this year it came back to within 100 meters of where it was last year." He records the bird's latitude and longitude. "These birds were coming here before there were landmarks, like these houses." He sees another piping plover whose presence he recorded for the first time in September 2004. "I've seen this bird over 200 times. There's something about seeing a bird that goes so very far away each year 200 times."

Calls start coming in. Birds are still falling prey to red tide. "We've got a red tide cormorant. Can't hold its head up. Limp. Can't fly. It's off the Horace Caldwell Pier in Port Aransas. We'll capture it if we can, but the double-crested cormorant is one bird you don't want to bite you. The hook on the end of its beak is powerful. Designed to hold onto squirming fish. It will tear you up." Amos is a self-taught birder. He's been at it a long time, starting years ago while working at the Lamont Doherty Earth Observatory at Columbia University. "That got me into birds, hour after hour on the deck of a ship, working the equipment, looking at birds, the sooty shear-water, the wandering albatross. I counted and drew them."

We find a dead brown pelican. Its legs are blue. Amos takes it into the dunes. Fog rolls in. People walk their dogs, or, rather, the dogs walk and their owners drive alongside. Amos takes another call. Someone has captured the cormorant. We're halfway through. "When I started, there were no houses here, no condominiums. I'd reach this point, sit on an old tree that had washed up in the dunes, and contemplate. One day someone took away my tree and put Gulf Shores Condo in its place. In 1980, Hurricane Allen engulfed them." He sees another piping plover he knows. "Hello, my dear." We're at the Corpus Christi city line, and the beach is thinning. "Ah, the lesser black-backed gull. It breeds primarily in Europe and Iceland. It's expanding its winter range. Once rare here, it's common now."

A dead fish, a red drum, has just washed up. "The red tide is still here." We see a huge plastic bag that once was filled with rice, then a big piece of plastic that once covered a pallet somewhere. "These big sheets of plastic get buried in the sand. They are impossible to remove. I don't have a category for this. I struggle to categorize all the plastic." Another call. The National Park Service has a green sea turtle. We'll meet at the bank to collect it.

A lot comes onto the beaches of Texas's barrier islands—Sandifer's boat, for example. Under previous ownership, coming in to unload a shipment of marijuana, it was seized by federal agents who arrested the crew (the driver escaped into the dunes), seized the cargo, and auctioned the boat, which Sandifer bought.

By the time we've gone seven miles, we've seen a lot of birds, a lot of beverage containers, and a lot of other plastic, but no knots. One year, Amos saw 1,600. Between 1979 and 2007 the number fell by 54 percent. At the same time, the number of people walking or driving or camping on the beach increased fivefold. Whether the decline of knots on Mustang Island reflects a loss in the Texas population or whether the birds fled the disturbance to find a quieter place is hard to say. "The Christmas count is on Monday. They always expect me to get a red knot, but I guarantee that this year I won't."

The green turtle is in sad shape. Fishing line is wrapped around its neck, and it seems to have swallowed, or tried to swallow, the hook. Its breath is raspy. Amos will have the turtle X-rayed to see if the hook has set and where. At ARK, Amos weighs the thin turtle, then puts him on a towel in a plastic yellow kiddie pool. The line is still around his neck. "I'll easily cut that. If the hook is in his throat, he can have surgery. If he's swallowed it, there's nothing we can do." We look at other sea turtles mangled by fishing lines and boat propellers.

Amos released five redheads yesterday. "I've seen so many redheads in trouble recently. They're emaciated. If you see dead birds on the road, they may be redheads. It's odd. I wouldn't think the red tide would affect them." So far this year, Amos has rehabilitated 1,400 birds. I see the cage where he kept the knots. "I put them in this big pen, covered it with cloth to keep them calm. They were low in weight. I fed them mealworms and chopped fish." When the birds recovered, Newstead released them. He hasn't yet resighted any of the rehabbed birds.

Newstead is still looking for knots. A few days later, driving out of Corpus Christi, he and I pass field after field of brown stubble—what's left from harvests of sorghum, corn, and cotton. Sandhill cranes are eating the stubble. In a wet year, rains would soak the fields, turning them into ephemeral ponds, attracting hundreds of roosting birds, but now the soil is bone dry. The landscape is empty. Land is for sale. A 175-turbine wind farm has been proposed for these fields.

In the blink of an eye, we pass through Bishop, a small town that manufactures more than 7 million pounds of Advil and Motrin every year. Before the day is out, I will need some. Newstead points out 25 curlew in a field, a prairie falcon gliding in and, flying by, a bird he sees as destined for the endangered species list, the mountain plover. We turn at a sign for the Bishop airport, park near its one runway. Anse Windham takes us to his plane. A small (24-foot) single-engine propeller plane, brightly painted red, white, and blue, a large lone star on the nose, it looks like an enormous Texas flag. A long antenna is attached to the underside. I scrunch into the backseat, put on headphones. We'll be in the plane most of the day, listening and looking for knots.

A little over two months ago, Newstead trapped 11 knots and glued radio transmitters between their shoulders, close to the skin. Each transmitter sends out signals for about three months, until the battery dies. Of the 11 radio tags, 10 still work. The eleventh bird had been emitting a signal from the same place for three weeks. Newstead found the bird, hundreds of yards back in the dunes, dead. Sometimes when the adhesive, a five-minute epoxy, deteriorates, the transmitter falls away, and Newstead finds it still pinging, the bird long gone. The 10 birds with working transmitters left the beach some time ago. Newstead is hoping the radio signals will tell him where they are. He turns on the receiver.

It's a clear day, and gusty. Four sandhill cranes are flying below us. We're heading out toward the sea, flying over swales normally filled with over two feet of water at this time of year but now empty. The plane bumps in the wind. We turn by the naval station to check on a piping plover—part of a different radio-tracking project—and then by Pelican Island, a 400-acre dredge pile. In the 1970s, DDT had whittled the pelican rookery down to five nesting pairs. Once the pesticide was banned, the Texas population rebounded to approximately 8,000 nesting pairs. Seven years ago, raccoons and feral hogs arrived on the island, forcing out the birds. We fly over a coyote stalking a spoonbill and 10,000 redheads in a pond.

Windham turns south, flying down Padre Island. The sea is to my left. If the water were clear, which it isn't, we'd be seeing dolphin and shark. Instead, it's stained with red tide. The beach is still strewn with dead fish and dead pelicans. To my right are the Padre Island dunes, and beyond them, the tidal flats and shallow waters of the Laguna Madre.

No rivers flow into the Texas laguna. Freshened only by rain and occasionally rain-filled transient creeks, laguna water is saltier than the sea. In the past, as water evaporated in the summer heat, salinities in the laguna would double, triple, quadruple those of the gulf. When the water turned to brine, fish died. In between the die-offs, the laguna's broad meadows of seagrass nourish an abundance of fish, its redfish and spotted seatrout legendary. The fish adapt to the laguna's extreme conditions, heading for deep water during the cold snaps and toward the passes when salinities soar. The laguna constitutes only 20 percent of Texas's bays, yet for many years produced 60 percent of its finfish.

We fly past Yarborough Pass, opened by hurricanes, now mostly filled in with deep, soft sand. Newstead released the rehabilitated knots here. Windham shields his eyes in the glare. It's getting warm. We're flying into a 30-knot headwind. About 60 miles down the beach, we turn west, cross the beach along the Mansfield Channel, and fly out over the laguna. Knots could be anywhere in the laguna's 600 square miles. "Sometimes I think this is a fool's mission, to look for small birds over such a vast area," Newstead says. "I can't walk the laguna, or even cover it by boat. It's just too big." Looking for radio-tagged birds by plane is one of his best options.

The human imprint is everywhere: in passes cut through the beach to the open sea; in the massive causeway built from Corpus Christi out to Padre Island; in the 12-foot-deep, more than 100-foot-wide shipping channel, the Gulf Intracoastal Waterway; and in the Land Cut. The laguna is also highly protected. The Padre Island National Seashore, the Atascosa National Wildlife Refuge, and mile after mile of the private King and Kenedy ranches border 70 percent of the laguna. Salty, seemingly inhospitable, its shallow waters host 79 percent of Texas's seagrasses, including shoalgrass, the favorite and pretty much only food of laguna redheads. The mesquite- and prickly-pear-covered islands of the laguna include several hundred artificial spoil islands built when the Gulf Intracoastal Waterway was dredged in the 1940s. They are rookeries for approximately 20,000 pairs of colonial waterbirds: black skimmers, seven species of tern, five of heron, four of egret, and two of ibis. Spoonbills nest here as well, along with brown pelicans and a rare coastal colony of American white pelicans. More than 2 million birds nest, winter, or migrate through the laguna.

Newstead directs Windham to the Laguna Atascosa National Wildlife Refuge, where he saw knots last week. He thinks they may winter in flats near the refuge. "Turn here," he instructs Windham. "This algal flat was mois- tened by rain last week. It's getting productive for shorebirds. It's between two and six inches deep. Good for red knots." We're getting radio signals, but they are faint. He compares radio telemetry to dowsing. "I think there's a knot!" Newstead nods toward the window. The plane swoops. He wants a better look. Windham obliges. We dive three or four more times. I'm getting dizzy and a little clammy. We don't find the knot. The plane levels off, and the radio signal kicks in again, sounding like a softly chirping cricket. We swoop low. Newstead and Windham work well together, clearly enjoy each other's company. They barely discuss strategy. Windham homes in on the signal; Newstead, peering down onto the flats, merely gestures or says "Here," and the plane swoops again. This efficiency allows many nosedives.

The plane circles low. It flushes a few birds, noses closer. Newstead, ever the optimist, gets excited. We've had more than a dozen possibilities today, but they haven't panned out. The transmitters have been on the birds for 60 days. People are easier to see. "There's the border patrol," he observes, "looking for walkers." We're so close to Mexico Windham checks in with Homeland Security. If he doesn't, he tells me, there are consequences, like being met by armed guards and searched when he returns. The border winds toward the sea, following the many and tight curves of the Rio Grande. Windham follows the radio signals while taking care to stay in the United States. We're approaching the border. He turns away at the light-house and flies over a large home, with a private runway, belonging to a duck hunter. I'm trying to listen as Newstead and Windham discuss the fine points of radio telemetry, but we've been diving and rolling all morning, and I'm about to throw up. We pick up the signal again. Windham makes a sharp turn. The signal grows stronger. We swoop. The bird stays hidden.

Wind jostles the plane. For most of the year along Texas's southern coast, it blows strong and steadily late in the afternoon, when electricity consumption peaks and prices rise. Texas is a leader in the wind energy field, generating 7.4 percent of its electricity from wind, and in the next few years, wind capacity along the coast is expected to double. A 300-turbine wind farm has been proposed off Padre Island. Sandifer has seen flocks of 500,000 black terns feeding on anchovy shoaling near the proposed sites. Newstead's not sure whether it's even possible to study avian mortality offshore, where finding dead birds to count will be next to impossible. Shawn Smallwood, an ecologist who studies avian mortality and wind farms in California, agrees. "There is no method available to estimate fatality rates offshore, and to my knowledge nobody has done it with any confidence," he wrote me.

It will become imperative to do so. Bryan Watts at the Center for Conservation Biology at the College of William and Mary finds that "build-out of the wind industry along the Atlantic Coast will result in the largest network of overwater hazards ever constructed." Many birds can sustain losses from offshore wind farms along the Atlantic, but Watts determined that vulnerable populations of knots, roseate terns, piping plovers, American oystercatchers, and marbled godwits may not. Joanna Burger from Rutgers University, evaluating the risk of offshore wind energy to threatened shore-birds, found that, for red knots, turbines sited in critical bays or estuaries where birds winter or refuel might pose a serious threat.

Newstead worries about large numbers of turbines in the middle of a bird migration highway. "East of the Rockies and west of the Mississippi River, birds traveling between South or Central America and northern breeding grounds funnel right through here," he says. Redheads, consuming saltwater every day in the laguna, retire to freshwater ponds to relieve the osmotic stress. The 260 turbines on the Kenedy Ranch stand between the laguna and a mosaic of freshwater ponds used by 80,000 redheads. More turbines are planned for an area near the Laguna Atascosa Refuge, home to over 400 species of birds, including endangered northern aplomado falcons. As wind technology improves, the turbines grow bigger—blades over 400 feet high sweep over an acre or an acre and a half, spinning at 140 miles per hour at their tips.

In our small plane, we're no longer getting a radio signal. Windham glides over another part of the flat. "We need to cut this short," he says. "This wind is using up my gas. We have enough left to go in one direction for fifteen more minutes—your pick. Then we turn back." Newstead doesn't want to leave, but he doesn't have much choice. He and Windham track one last soft signal, and then we head north, passing the wind farm at the Kenedy Ranch.

Data on avian mortality, lacking for offshore turbines, exist for some terrestrial wind farms. As the number of turbines rise and the calculations of avian mortality grow more sophisticated, the death increase. Over the years, estimates made by the Fish and Wildlife Service grew from 33,000 birds dying from wind turbines per year to 440,000. The latest calculation, in 2013 from Smallwood, raises the number to 570,000. Texas ranks first in the country, as of 2013, for number of megawatts installed, and second for the number of utility-scale, large turbines (more than 7,500). Six of the country's 10 largest wind farms are in Texas, and of all the states, Texas added the most capacity in 2012 (1,800 megawatts). Most Texas reports of bird fatalities from wind turbines are confidential: Smallwood could not include them in his calculations.

The Kenedy wind farm has high-tech radar that, when it senses masses of birds approaching in storms or fog, can signal the turbines to shut down until the flock has passed, offering the potential to avert fatalities in flocks of migrating birds. But because the data from these kinds of systems, used in other wind farms as well, have not been made public, both Smallwood and Edward Arnett, who studies bat mortality from wind turbines, say such systems haven't yet proven themselves. On a clear, sunny day, Newstead, working at the Land Cut, watched a turbine blade smash a white pelican as its small flock approached the towers. He doesn't know whether he witnessed an isolated incident or something that occurs regularly. Avian mortality data in Texas are infrequently peer reviewed, and not available to the public. In addition, Smallwood believes that the industry may not be looking for dead birds often enough, long enough, or in a wide enough sweep to fully estimate how many birds die, especially small ones.

Wind power can reduce U.S. consumption of fossil fuels, substantially reduce carbon dioxide emissions and, according to the U.S. Department of Energy, supply at least 20 percent, and perhaps more, of the U.S. demand for electricity. By the end of 2012, 45,000 wind turbines were operating in the United States, a number expected to triple by 2020. The number of bird fatalities caused by wind turbines is small compared to the hundreds of millions of birds that die colliding with buildings and windows each year, the 1 to 4 billion birds that cats kill annually, the 5 to 7 million killed by communication towers, the 60 to 80 million killed by cars, or the 70 to 90 million killed by agricultural pesticides. Still, wind farms can be built to reduce the danger to birds.

In California's Altamont Pass, wind farms killed thousands of birds, including golden eagles, kestrels, burrowing owls, and red-tailed hawks. Removing turbines sited in avian "hot spots," replacing older ones, and shutting them down in winter halved the number of eagle deaths. In Spain's Cadiz province, 300 wind turbines stand along the autumn migration route of griffon vultures crossing the Strait of Gibraltar into Africa. Scientists observing griffons flying dangerously near turbines alerted the control towers, whose workers turned them off until the birds had safely passed, reducing mortality by 50 percent.

Our plane lands. It's not been a great day for knots. Windham and Newstead disembark, discovering that a piece of antenna is now missing, possibly sheared off in the wind. They joke about how, given the wind speeds, they ought to tighten the screws holding the plane together. The next time Newstead flies, a few days later, he won't find many knots either. Tracking birds demands a high degree of patience and persistence, and that Newstead has. Eventually, as Windham and he continue to fly, he will find knots, 40 or 50 near Baffin Bay, an inlet of the laguna, and 500–600 in flats near the Laguna Atascosa National Wildlife Refuge. Data loggers will yield even more information.

———————————

By the start of 2014, Newstead, working with Larry Niles, had retrieved about one-quarter of the approximately 100 geolocators they'd deployed in Texas. The Texas data disproved the notion, jokingly aired, that Newstead's failure to see flagged birds in Texas resulted from not cleaning his spotting scope often enough. The data confirmed that most Texas red knots winter on the Gulf Coast, spending more than nine months of the year there. One juvenile, born the year it was tagged, spent its first summer in Texas.

Newstead now knows that even when he can't see birds himself, they are somewhere in the gulf's bays and lagoons, between Mexico's Laguna Madre de Tamaulipas and Texas's Laguna Madre, and perhaps as far north as Matagorda Island. His aerial surveys suggest that the number of knots wintering in Texas, once thought to be about 300, might actually be much closer to Sandifer's observations of several thousand. The data loggers Niles and he deployed held other surprises. Not only did they reveal that adult and juvenile knots winter in Texas, they hinted that knots were still taking an old, once-well-traveled route north—a route that went overland, far from the sea.

In 1912, Wells W. Cooke of the U.S. Biological Survey recalled an "almost endless succession" of shorebirds migrating across the prairies of Kansas, Nebraska, and the Dakotas, and that birds traveling up through the prairies of the Mississippi valley formed "the great highway of spring migration." Knots, among the shorebirds taking that route, were "tolerably common." Even though they were a decreased and "diminutive army" when Forbush is writing, in 1912, knots traveled in "numbers"—he doesn't say how many—from Texas up through the Mississippi valley. Today, that route is quiet for knots, most of the birds killed by hunters long ago. A 1958 record of the red knot in Texas describes it as a casual winter visitor migrating through the Great Lakes region.

The U.S. Geological Survey mapped what remains of the Great Plains route, collecting sightings over a 10-year period between 1986 and 1995, finding knots along the Texas coast from Boca Chica Beach near the Mexico border up through Padre and Mustang islands, sand flats at the Port Aransas Airport (2,500 knots sighted by Amos), Matagorda Island, and the Bolivar Flats near Galveston. The survey also recorded them farther north: some 19,000 birds at Last Mountain Lake, the Quill and Chaplin lakes in Saskatchewan, Beaverhill Lake in Alberta, and in a few areas in the United

States—Utah's Great Salt Lake, the Arkansas River in Kansas, Oklahoma's Cheyenne Bottoms National Wildlife Refuge, and the Oologah Reservoir, Tulsa's water supply.

Some of these stops are prairie potholes—spongy depressions left in the ground thousands of years ago as North America's glaciers receded. Small, shallow, occasionally salty, they lie scattered across the Great Plains of the United States and Canada. Texas red knots, along with thousands of other shorebirds on their way north from Texas, replenish their energy reserves at prairie pothole lakes.

The lakes, sculpted by glaciers, are being redesigned by humans. A large deposit of sodium phosphate extracted for detergents, livestock feed, starch, carpet deodorizers, textiles, and the paper industry underlies long, narrow, Chaplin Lake. Saskatchewan Mining and Minerals, which controls the lake's water levels for brining, helps maintain habitat for shorebirds. At the Quill Lakes, Ducks Unlimited regulates water levels, enhancing habitat for ducks and shorebirds. In the early 1990s, the lakes attracted as many as 9,000 knots. No one knows what they eat there to regain their strength. A third of the continent's sanderlings refuel at Chaplin, Old Wives, and Reed lakes, filling up on flies and brine shrimp. At the Quill Lakes, dowitchers and godwits eat midges, pondweed, and occasionally grasshoppers. Perhaps knots share this repast, perhaps not.

When rain and melting snow inundate the lakes and nearby pastures, shorelines disappear. Knots, at their peril, may be forced to roost on the nearby highway, where one foggy night in 2011, 10 were reported killed near Reed Lake. When the lakes are severely flooded, some birds pass over them, continuing on to the mouth of the Nelson River on Hudson Bay, a previously unrecognized stopover.

State birding organizations tracking and confirming rare sightings and eBird, the world's largest electronic database of bird sightings, trace additional stops on this interior route that red knots still travel. Birders tell me they still see red knots, usually only one or two, not only along Oklahoma's Oologah Reservoir but also at Lake Hefner, dammed to create water for Oklahoma City; at the rare inland salt marshes of Kansas's Quivira National Wildlife Refuge, near the great bend in the Arkansas River; in the shallows of Stone Lake, a prairie pothole in South Dakota (where a lone knot was spotted); in the prairie potholes of North Dakota as well as the sewage lagoons of Grand Forks; and on Lake Oahe, a reservoir built behind a dam on the Missouri, also in North Dakota, where in 1977, a flock of 40 birds was sighted.

Ducks Unlimited, working with other conservation organizations and the U.S. Fish and Wildlife Service, is preserving what's left of this old route. Millions of ducks and geese pass through or nest in the prairie potholes of the Great Plains: the Duck Factory is one of the most important breeding grounds for waterfowl in the world and, in North America, one of the most threatened, as acre after acre of grassland and wetland is turned into fields of soybeans and corn. Ducks Unlimited estimates that between 50 percent and 90 percent of prairie potholes have been degraded. Scientists calculate that those remaining are worth more—$4 billion more—for their wildlife, flood and erosion control, and ability to store carbon, than they would be as agricultural fields. A top priority for Ducks Unlimited is raising the money to conserve 2 million acres of grassland and to protect and restore 400 shallow lakes.

Shorebirds may benefit. Most red knot sightings in North Dakota come from the region of prairie potholes. Some birding guides rate the chances that birders will spot particular species. From North Dakota, Dan Svingen with the U.S. Forest Service and Lawrence Igl with the USGS Northern Prairie Wildlife Research Center sent me a ranking for red knots—"How Lucky Can You Get." The sightings are infrequent: a knot here, a knot there, and once in awhile perhaps a larger flock, each sighting an inkling of what might have been, and still could be.

Ten

DOES LOSING ONE MORE BIRD MATTER?

The knot ghost trail from Texas up through the Great Plains, once populated but now rarely used and fading from memory, is disquieting. What might it mean if knots, whose numbers have already declined substantially, are lost from the edge of the sea?

Each spring in my backyard, when the ground is still bare and brown, a shorebird calls for a mate. A brook runs along the meadow, freshwater lacking a sandy shore, too far upstream to feel the tide. When dusk falls and a few stars appear and other birds cease to sing, we would hear the call of woodcocks—shorebirds that evolved with sandpipers but left the open beach for a life inland. I rarely saw these reclusive, well-hidden birds on the ground: their low, nasal *peent, peent, peent* located them. After calling for a minute or two, a male would take off—once within a few feet of where I was standing, coming directly toward me, a surprise, I think, to both of us. He'd then shoot into the sky, spiraling beyond the tree line, out of sight. A fluttering sound—air rushing through his tail feathers—signaled his return. He executed careening barrel rolls and loops, accompanied by melodious warbling, before dropping precipitously to the darkened ground, camouflaged in the dead leaves. *Peent, peent, peent!* He'd safely landed. He repeated his aerial dance a bit later each evening as the days grew longer.

What of this decidedly not-of-the-shore shorebird? Over the years our singing field grew quieter, down to one woodcock barreling his way through the sky, performing for a female I hoped was waiting in the darkness. It didn't take much to send the others away. Two new houses and their resident dogs, running and barking in the field, interrupted the nuptial displays. Now, in early summer evenings, I either listen for the lone bird, walk down the road where flights are still going strong, or drive across town to the seine field, where fishermen once mended and dried their nets by day and woodcock still sing in the evenings. Their disappearance from our meadow, though sad, isn't particularly significant in the overall well-being of woodcock: feeling pinched, perhaps they had better choices. In the United States, loss of young forest led to population declines in the 1970s and 1980s, but in the past 15 years their numbers have held steady.

The same can't be said for their sand-loving cousins. Each lineage of knot is declining. Ruddy turnstones along the East Coast of North America have dropped by 75 percent since 1974. Along the East Coast of North and South America there are fewer semipalmated sandpipers: fewer in Delaware Bay; fewer in southern Ontario and in the great tidal flats of the Bay of Fundy, where they double their weight on tiny amphipods scavenging in the mud before heading south; fewer in their winter homes in the Guianas, where their numbers have dropped dramatically—79 percent since the 1980s; and no more breeding birds in Churchill, Manitoba, on the shore of Hudson Bay, where in the 1940s the most abundant breeding sandpiper was the semipalmated.

Other North American shorebirds—killdeer, lesser yellowlegs, and whimbrel—are also declining. Each year elegant whimbrel fly from their breeding grounds in the MacKenzie River delta to their winter homes in South America, pausing on Virginia's barrier islands, where, between 1994 and 2009, they've declined by half. I had believed that birds I see on the beach each year—handsome black-bellied plovers with their tuxedo plumage, semipalmated plovers with their black collars, sanderlings arcing through the sky—were an immutable part of my seascape. They're not. Each spring across the world, 50 million shorebirds fly north to breed in the Arctic. Where trends are known, almost half their populations are declining. In the Canadian Arctic, shorebirds are declining much more rapidly than other birds, their populations dropping by as much as 60 percent. In the

United States, half the birds that depend on beaches, salt marshes, estuaries, and tidal flats on the coast are declining. In Europe, 12 million birds breed, molt, refuel, or winter in the Wadden Sea, Europe's great crossroad for migratory birds. Twice as many waterbirds are in decline here (41 percent) as are increasing (22 percent). All over the world, shorebirds are losing their homes.

In *A Sand County Almanac* Aldo Leopold said of the woodcock: "The woodcock is a living refutation of the theory that the utility of a game bird is to serve as a target, or to pose gracefully on a slice of toast. No one would rather hunt woodcock in October than I, but since learning of the sky dance I find myself calling one or two birds enough. I must be sure that, come April, there be no dearth of dancers in the sunset sky." Woodcock hurling themselves into the twilight; thousands of chattering shorebirds suddenly sweeping into the sky at dusk on a spring evening; sharing the autumn marsh with a young whimbrel finding its way, without its parents, to a home it's never seen and, like the knots, flying thousands of miles each year; a bar-tailed godwit, perhaps the most extreme long-distance flier of them all, traveling a record-breaking 7,200 miles nonstop, from the pristine barrier beaches of Alaska's Yukon River delta to its winter home in New Zealand: if all these birds were gone, my world would be sadly diminished. But does that matter?

The foundation of food webs may not be apparent until they fray, the value of individual threads unseen until the fabric is torn. The loss of horseshoe crabs, decimated in the nineteenth century for fertilizer and again in the twentieth for bait, reverberates on shore and in the sea. Shorebirds feel the loss, both as their seaside homes deteriorate and as their energy-rich food of horseshoe crab eggs declines. Shellfishermen feel the loss of horseshoe crabs as well. In Massachusetts, horseshoe crabs, considered vermin that preyed upon commercially valuable clams, were killed, up to 1 million each year. Shellfishermen were paid bounties for horseshoe crab tails, and in eight towns they were legally required, as recently as 2000, to kill every horseshoe crab they found. Now that far fewer rototill and aerate the sand for oysters and quahogs, and far fewer break up thick colonies of bamboo worms that prevent seed clams from settling, shellfishermen are asking regulators to ban the bait industry in Wellfleet Bay so that horseshoe crabs can return.

Loggerhead sea turtles feel the loss of horseshoe crabs. Born on the beaches of Florida, they ride the Gulf Stream across the sea in their youth, growing up off the Azores before journeying back to lay their eggs on the sands of their birth. On the journey, they forage along the eastern seaboard. In the 1990s, as bait fishermen began removing hundreds of thousands of horseshoe crab from coastal waters along Virginia and Maryland, loggerheads, losing their preferred prey, turned to blue crab.

As blue crab declined, hungry loggerheads, with fewer choices, began eating croaker, a grunting, shallow-water fish, and small menhaden. Fish easily outswim and outmaneuver the 250-pound sea turtles; loggerheads began eating live fish from fishing nets or fish freshly discarded by fishermen. In the lower Chesapeake, the number of sea turtles has declined by 75 percent since the 1980s. Pelagic long lines and shrimp nets continue to hook and trap thousands of loggerheads, and now they may also lack essential food.

Sea turtles are vital to a healthy ocean. When Columbus sailed to America, as many as 600 million green sea turtles plied the waters of the Caribbean, so many, according to a 1774 history of Jamaica, that "vessels, which have lost their latitude in hazy weather, have steered entirely by the noise which these creatures make in swimming, to attain the Cayman Isles." The sea is quieter now. To some, sea turtles may be odd anachronisms, relics of a distant past, but their diminished numbers have left a mark on coral reefs now covered with algae, and in dying sea grass that once sheltered young fish. Sea turtles, especially leatherbacks, eat hundreds of pounds of jellyfish each day. Their loss, along with the depletion of once-large fisheries, leaves a sea of jellies with few predators. Nesting loggerheads help stabilize sand dunes. Addled eggs decompose on the beach, supplying nutrients that encourage vegetation on dunes increasingly vulnerable to erosion.

What about birds? Where do they fit in? In the bush of northern Kenya, honey collectors whistle through their palms, snail shells, or hollow date nuts to call a bird, the greater honeyguide. The white-breasted birds respond. Calling, flying short distances, and then perching, they lead humans to hives located in trees, rock crevices, and termite mounds. The honey gatherers then smoke out the hives, quiet the bees, and collect the honey. The birds eat larvae and honeycomb left in the hive. Each needs the other. Humans, following the birds, find hives in one-third the time. Honeyguides can't get into 96 percent of the hives without people or other large animals to open them, an explicit, visible manifestation of an essential connection between humans and birds that enhances the lives of both. Today, scientists are beginning to articulate other ways—economic and ecological—in which our lives are enhanced by the presence of birds and diminished by their absence.

An early impetus to protect birds was economic. Scientists in the Department of Agriculture's Division of Economic Ornithology, founded in 1885, and its successor, the Biological Survey—where Charles Sperry analyzed the stomach contents of shorebirds—fought to prevent game hunters from extirpating shorebirds whose presence benefited farmers financially. Crop-devouring insects were eating 10 to 20 percent of the country's fruit, cotton, and cereal crops. Shorebirds ate insects. In 1911, the survey's Waldo L. McAtee wrote in an agency circular, "The protection of shorebirds need not be based solely on esthetic or sentimental grounds, for few groups of birds more thoroughly deserve protection from an economic standpoint. . . . So great, indeed, is their economic value, that their retention on the game list and their destruction by sportsmen is a serious loss to agriculture." Shorebirds, he wrote, "feed upon many of the worst enemies of agriculture."

Killdeer and spotted sandpipers, he found, ate army worms, animals that in 2012 laid waste to western New York's alfalfa and corn fields. Of crane flies and their larvae, eating their way through wheat fields, McAtee writes that "among their numerous bird enemies, shorebirds rank high," noting that phalaropes, woodcock, pectoral and Baird's sandpipers, plover, and killdeer all feed on them. Scientists working for the Biological Survey cut open thousands of birds to determine what they ate and how much. Grasshoppers, McAtee found, were a staple of 17 species of shorebirds, including knots, dowitchers, plovers, and curlews. Shorebirds ate cutworms and tomato worms, weevils, and beetles. Knots, along with other shorebirds, ate crayfish, a pest in southern rice and corn fields, and sandworms, which preyed upon tiny oysters.

For many summers my neighbor and I shared a large garden, yielding a generous harvest of summer vegetables, with leeks and beans and peas to share with friends and tomatoes to can, until one summer when, despite years of conscientious companion planting and crop rotation, composting, and application of organic fertilizers, we lost almost everything. Pests devoured cucumbers, Brussels sprouts, broccoli, winter squash, and watermelon—everything but hot peppers. We picked at borers, beetles, and flies, brought in hungry nematodes, and planted again, with mixed success. Over 100 years ago, one H. W. Tinkham from Fall River, Massachusetts, watched spotted sandpipers cleaning cutworms, cabbage worms, and squash bugs from his garden throughout the summer. Until I read about Tinkham, I'd no idea what I was missing.

Today, ecologists are again, but more precisely, quantifying avian benefits to agriculture. The larvae of codling moths make apples wormy: birds take out 90 percent of their cocoons. Bird nesting boxes introduced into Dutch apple orchards reduced caterpillar infestations and increased yields by 66 percent. Ducks feeding in rice fields thin out, by 80 percent, the golden apple snail, a voracious pest. On coffee plantations in the Blue Mountains of Jamaica, birds forage on coffee borers, the coffee bean's most insidious pest, capable of destroying 75 percent of a crop. In Costa Rica, birds reduced borer infestations by 50 percent, increasing the value of the yield on each coffee plantation by an amount equal to Costa Rica's average annual per capita income. On shade-grown coffee plantations, where birds could roost in nearby forests, there were more birds and less borer infestation. In Canada's boreal forests, the financial value of birds controlling pests is worth $5.4 billion per year. The scientists who worked for the Biological Survey strongly believed that shorebirds reduced pests in agricultural fields. With so many fewer shorebirds, it's difficult to calculate what they could contribute today if their numbers were restored.

Pesticides also control pests, and their advent ended the economic ornithological work of the Biological Survey (later reconstituted as the U.S. Fish and Wildlife Service). Whatever the economic benefit of birds to agriculture, the benefit of pesticides was considered greater: a $10 billion investment in pesticides increased crop yields by $40 billion. David Pimentel calculated the cost in human health effects—cancers, respiratory, neurological, and cognitive impairments, pesticide residues in food, domestic animal poisonings, loss of beneficial natural predators, pesticide resistance, losses of honeybee pollinators, pesticide contamination of crops, ground and surface water contamination, and birds, mammals, and other wildlife poisoned and killed—to be about $12 billion a year, a cost paid by the public and by wildlife, and possibly underestimated by 100 percent. In this case, an economic valuation of nature didn't tip the scales on nature's behalf.

More than 50 years after Rachel Carson wrote *Silent Spring*, pesticides continue to kill at least 67–90 million birds in the United States each year. The American Bird Conservancy, concerned that today's best-selling and most widely used pesticides, nicotinelike neonicotinoids, are killing honeybees and birds, has asked the EPA to ban the use of neonicotinoid-treated seeds and to suspend other uses until their effects on wildlife have been fully tested. According to a 2013 review undertaken by scientists for the American Bird Conservancy, a single seed of neonicotinoid-treated corn can kill a songbird, and far lower exposures can inhibit their reproduction. In the Netherlands, on farmland where water concentrations of one neonicotinoid exceed .00002 parts per million—or greater than one drop of water in an Olympic-sized swimming pool—songbirds are declining at 3.5 percent per year. In the United States, USGS scientists detected the same neonicotinoid at more than twice that concentration in midwestern streams when rain falls during the planting season. Today, scientists find that toxic pesticides are four times more likely to be associated with the decline of grassland birds than is the loss of their homes. We have already experienced one silent spring. It is almost inconceivable that once again, we may invite another.

One November day in 1963, fishermen off the southern coast of Iceland watched as the sea turned brown, smoke shot from the water, and exploding magma rocked their boats. In the rain of ash and lava, Surtsey, a new island named for the black fire giant of Norse mythology, rose from the sea. The eruption continued for more than three years. Scientists couldn't land on the scorching shore. When the lava cooled, the barren rock began to green. Seaweed grew along the shore, moss at the crater's edge. Currents carried in floating seeds of rocket and sandwort, wind blew in seeds of willow, spores of lichen, fern, and moss. Birds too carried in seed: crowberry, buttercup, dock, meadow grass. In the 50 years since Surtsey was born, 70 species of flowering, seed-bearing plants took root and persisted: 75 percent arrived with birds. Closer to home, birds carried in seeds of huckleberries, blackberries, elderberries, strawberries, and mountain ash that are helping recolonize Mount St. Helens after the explosive volcanic eruption in 1980 destroyed much of the surrounding landscape. "Perhaps the least appreciated contribution of avian seed dispersers," biologist Çağan Şekercioğlu writes, "is enabling the colonization and regeneration of barren, deforested, ephemeral, remote, post-glacial, volcanic, and other marginal habitats."

More than 150 years ago, Darwin sowed the seed of the idea of birds dispersing plants across the world. He picked seeds from bird excrement, planted them in his garden, and watched them germinate. Remembering that gales can blow birds—and the seeds they carry in their crops and gizzards, or on their feet—across the ocean, he wrote in *On the Origin of Species* that "living birds can hardly fail to be highly effective agents in the transportation of seeds." Darwin's idea still holds.

Today, scientists estimate that 33 percent of bird species disperse seeds. Large mammals once distributed seeds as they roamed tropical forests. With so many gone, birds play an increasingly crucial role. In Panama's rain forest, nutmeg seeds falling to the ground and germinating beneath the parent tree almost always succumb to weevil infestations. Brightly colored toucans swallow nutmegs whole, fly off, and then spit out the seeds, enabling nutmeg trees to proliferate. Similarly, there'd be no mistletoe without birds carrying this parasite to a living branch where it can take root.

The 33 percent estimate of seed-dispersing birds doesn't include ducks, shorebirds, and other aquatic birds. Yellow-legged gulls, eating the translucent fruit of a small African shrub, spread its seeds across the Canary Islands. Andy Green and his colleagues are carefully documenting the underestimated, overlooked role of shorebirds in transporting seeds over long distances. Catching birds in the brackish ponds and marshes of Doñana, in the delta of the Guadalquivir River in southwest Spain, they found salt marsh seeds in the feathers of wintering redshank, a red-legged sandpiper. In the nearby Odiel marshes and salt pans they, like Darwin, collected bird excrement and culled the seeds. They planted iceplant, sowthistle, and glasswort seeds from the droppings of redshank and godwit, birds that fly hundreds of miles between stopovers and that fly nonstop between Odiel and northern Europe. Between 45 and 76 percent germinated.

They are not the only shorebirds eating seeds. In the salt marshes of Spain's Cádiz Bay, the droppings of wintering stilt and curlew also harbor seeds of salt marsh plants. Godwit in France's La Camargue, avocet, dowitcher, least and western sandpiper in the Texas Playa Lakes, and knot and curlew sandpiper in the coastal lagoons of Ghana all eat seeds. Each year, 1.5 million shorebirds, including some 700,000 redshank and godwit, migrate along the East Atlantic Flyway between Africa and the Arctic, perhaps dispersing plants along the edges of entire continents.

As the last Ice Age ended, birds seeded vast expanses of emerging mud and rock, turning them to lush salt marsh. Today, as the climate warms and the water rises, some marshes won't be able to keep pace, but seed-carrying shorebirds may save some of them: a curlew or a knot or a stilt, dropping a seed or two or three or many a bit inland, may give the grasses a fresh start.

Fox, coyote, fisher, deer, and wild turkey all regularly come through our meadow. One day, as a mother turkey and four babies meandered across the field, a coyote bounded from the tall grass and snatched one. Predator and prey disappeared in the blink of an eye, the only evidence of skirmish the frantic mother and her scattered chicks. Evidence of predation dissipates quickly. Another day, a deer was killed in the meadow. Within minutes, turkey vultures arrived. Within hours, the crows and they had picked the bones clean.

In traditional Zoroastrian funeral ceremonies in India, Parsis may lay their dead on the roofs of stone "Towers of Silence," where vultures quickly consume the exposed flesh. To the Parsis, vultures provide "great service to mankind in keeping clean the environments." In Tibetan sky burials, once the spirit has left a body, it too is brought to a mountaintop and given to the birds. Vultures serve others as well, in ways that weren't appreciated until the birds were killed.

In India, the anti-inflammatory drug diclofenac, used to treat lame cattle, threatened the very existence of vultures, killing, between 1992 and 2007, when it was banned, 96.6 percent and 99.9 percent of two species. Their unprecedented rapid, widespread, and lethal poisoning was a tragedy with unanticipated and expensive consequences. Without vultures to clean the carcasses of dead animals, the population of feral dogs exploded, their numbers increasing by 95 percent in one carcass dump in Rajasthan. India has the world's highest rate of human rabies infections; dog bites cause 96 percent of the deaths. Vultures not only protected people from rabies; their highly acidic secretions killed anthrax, brucellosis, and tuberculosis bacteria that infected rotting carcasses. The cost of bringing vultures to the edge of extinction—to human health, to tanners and bone collectors whose income disappeared, to communities now paying to dispose of accumulating carcasses—was almost $2.5 billion a year. Disregarding the near decimation of tens of millions of vultures in India from diclofenac contamination, Spain licensed the sale of the drug to treat pigs and cattle in the spring of 2013, putting its own vultures and eagles in harm's way.

India is not the only country to pay dearly for eliminating a bird. The United States may also be dealing with the legacy of extinction. The meadow and woods behind our house are exciting places for children to play, with secret passages leading through thick understory, a brook whose upper reaches flow from a hidden woodland pond, an old stone cow run bordered with blueberry bushes, and quiet hideaways lined with soft moss. In the midst of it all are disease-bearing ticks, some no larger than poppy seeds. Our children didn't contract Lyme disease while they were growing up, but by the time we'd lived here 30 years, the bull's-eye rash, fever, and chills associated with Lyme disease were familiar in the neighborhood. A short course of antibiotics quickly cured almost everyone, but others were hospitalized or incapacitated with joint and muscle inflammation.

Larval ticks hatch in the spring, disease-free, but acquire the Lyme spirochete when they latch onto white-footed field mice or chipmunks to take their first blood meal. The larger and by-then infected nymphs go on to feed on larger animals—people or deer—and pass on the disease. The cycle of transmission from small mammals through tiny, hard-bodied ticks to larger deer and people intensifies every two to five years, when oak and beech trees deliver a bumper crop of nuts, enabling the numbers of mice and chipmunks—and the ticks that feed on them—to skyrocket. The Centers for Disease Control estimates that 300,000 Americans are diagnosed with Lyme disease every year at an annual cost—in physician visits, blood tests, hospitalizations, medical procedures, prescription and nonprescription drugs, and loss of earned income—of $2.5 billion.

Human population growth and expansion into woodlands and meadows helps explain the surge in Lyme disease over the last decades. In addition, mice and chipmunk carriers may not have always experienced such dramatic population explosions. Passenger pigeons, once numbering between 3 and 5 billion, and constituting perhaps as much as 25 percent of America's birds, may have helped keep them in check. David Blockstein, senior scientist at the National Council for Science and the Environment, suggests that acorn-ravaging passenger pigeons, competing with mice for nuts—eating so many acorns that feral hogs went hungry—may have tamped down population surges, suppressing ticks and preventing the spread of Lyme disease. Eliminating the passenger pigeon, shooting entire flocks in their roosts and clearing their forest homes, could we have anticipated the rise of Lyme disease? If we lose shorebirds, it is our children who may pay, in ways we can't imagine or predict.

In 1999, West Nile virus appeared for the first time in New York City, perhaps from an infected mosquito coming in by airplane. West Nile virus has now spread throughout North America, within 10 years infecting 1.8 million people, killing 1,000, causing 13,000 cases of encephalitis or meningitis, and killing millions of birds. In the parts of the country most affected, numbers of American crows declined by up to 45 percent. As summer draws to a close, dead crows and blue jays signal the virus's return. I find their bodies in the yard and in the woods and fields where I walk, alerting me to retreat at dusk. West Nile virus is spreading in the United States, but not uniformly. Some birds don't transmit the disease as well as others: where a greater variety of birds graces the landscape, human incidence is lower.

Avian influenza viruses (AIV) pose a disease risk to humans and to billions of industrially raised poultry. If strains of AIV, harmless in wild ducks, seagulls, and shorebirds, spread to crowded chicken houses, the genes can rapidly reshuffle into a deadly combination which can kill or require preemptive euthanization of millions of chickens. Scientists believe that a 2007 outbreak on a poultry farm in Saskatchewan occurred when the farm, which usually uses municipal water, pulled water from a nearby pond without treating it, allowing introduction of a low pathogen strain of AIV from a wild duck. In the chickens, its genes recombined into a fatal virus.

Delaware Bay, where thousands of staging shorebirds and thousands of pairs of breeding herring and laughing gulls mingle on the beaches, is a "hot spot" for avian flu. There, researchers isolate AIV in shorebirds and gulls 17 times more frequently than at all other monitored sites in the world combined. The highest prevalence occurs in the spring among birds concentrated on or near Reeds Beach. David Stallknecht from the Southeastern Cooperative Wildlife Disease Study at the University of Georgia's College of Veterinary Medicine describes it this way: "Delaware Bay is unique—in no other place in the world do we see this diversity and prevalence of influenza in shorebirds and it occurs every year."

Red knots in Delaware Bay are rarely infected. Ruddy turnstones have the highest infection rates among shorebirds—11 percent versus .5 percent for other species. It's not clear why: when they arrive in the bay, they show "near-zero prevalence" of avian flu. Whether they are infected by ducks or gulls in the bay when they arrive or by some other means, every year the epidemic runs its course and subsides with seemingly little consequence. In Delaware Bay, researchers have a unique opportunity to gain insight into the transmission and circulation of AIV, and with that knowledge, help protect people and commercial poultry in the United States. In 2006, scientists isolated, from shorebirds and gulls, a nonpathogenic version of the AIV strain that killed Canadian chickens in a 2004 outbreak, providing, according to Stallknecht, a "heads-up" to local poultry producers.

They also isolated and analyzed avian flu gene sequences in gulls and shorebirds to help track another strain, H5N1, which caused major outbreaks in Asian poultry and infected people there. This analysis determined that it was highly unlikely migrating birds would carry this strain into North America, a great relief. As epidemiologists seek to understand how avian influenza moves from country to country and continent to continent and to assess risk, Delaware Bay provides an important laboratory to find answers. Ongoing shorebird and gull surveillance in Delaware Bay will continue to have implications: AIV prevalence in turnstones migrating through Delaware Bay is gradually increasing (even though their numbers have been declining), and is predicted to increase further as the Earth warms and the birds' annual arrival in the bay further overlaps with peaks of AIV in resident ducks.

In considering how the birds of Delaware Bay came to experience uniquely high rates of avian flu, scientists suggest that "dense aggregations of animals at dwindling stopover sites might create ecological hot spots for pathogen transmission among wildlife species," citing Delaware Bay as an example. The decline of horseshoe crabs, and the erosion of spawning beaches, crowded birds onto the few beaches where horseshoe crab eggs were still abundant. Benign strains of avian flu in Delaware Bay's ducks, gulls, and shorebirds haven't jumped into the nearby chicken farms of Delaware, whose Sussex County ranks as the nation's top-producing poultry county, selling millions of broilers every year. It would be a tragic and unanticipated consequence if diminishing the seaside homes of shorebirds, taking away their food, and altering the timing of their migration increases the prevalence of avian flu and its chances of spilling over into domestic poultry. In Delaware Bay, horseshoe crabs, shorebirds, and beaches need space.

The cool, nutrient-rich waters of the Humboldt Current abound with tiny marine plants that feed vast shoals of anchovy. At one time millions of tons of anchovy fed 60 million seabirds: cormorants, pelicans, and boobies that nested on the dry Chincha Islands off Peru's coast. Over hundreds of years, their guano piled up into mountains 150 feet high. High in soluble nitrate and phosphate, the bird droppings, ignored by Spanish conquistadores seeking South America's gold and silver, made excellent fertilizer. In 1858, at the peak of Peru's guano age, over 300,000 tons of excrement, mined by convicts and indentured servants from China, were exported to Britain. American farmers in Delaware Bay were fertilizing with horseshoe crabs, but before guano, British farmers used insoluble, and therefore highly unsatisfactory, bones to fertilize their fields.

Guano sales financed Peru's debt until the War of the Pacific, also known as the Guano Wars, during which Bolivia and its ally Peru fought Chile over bird droppings. The story of Peruvian guano didn't end with the Guano Wars or the introduction of synthetic fertilizer. Demand is rising again in a growing market for organic fertilizers, but now little Peruvian guano is left; about 4 million birds produce about 12,000 tons of guano a year. When anchovies were overfished, seabird populations crashed. A sea diminished by overfishing supports fewer birds and less abundance.

The loss of a bird can reverberate through a food web, touching its many strands in ways we have only begun to measure. Biologist Douglas McCauley's research on remote Palmyra Atoll reveals how disrupting one layer in a food web cascades throughout. Palmyra Atoll, a string of coral reefs and tiny islands halfway across the Pacific, is the only seabird nesting site within 450,000 square miles of ocean, and home to the world's second-largest colony of red-footed boobies—more than 6,000 pairs. Today, the atoll's native forests of large-leaved bougainvillea and beach heliotrope are broken by large stands of coconut palms propagated by commercial growers to produce coconut oil. The birds reject the palms and instead inhabit the native trees, where their numbers are five times higher. The area around the palms is diminished as a result.

Native soil, fertilized by boobies that have fed on squid and flying fish out in the open ocean, is five times richer in nitrogen than soil around coconut palms. Water running from native forest into the sea carries 26 times more nitrogen than water from the palms. Delivered on an outgoing tide or with the rain, it fertilizes coastal waters, yielding a greater abundance of tiny phytoplankton. They in turn feed more and larger zooplankton, which attract a greater number of manta rays swimming in to feed. Red-footed boobies nesting in Palmyra's native trees trigger one of the longer ecological cascades so far reported in nature, redistributing energy and nutrients from the ocean into the impoverished soil of the islands, and then back into the sea, where it moves from the tiniest prey through to large predators. Human introduction of coconut palms weakened and compromised the entire cycle. Plunge-diving red-footed boobies spend most of their lives at sea, but while they nest on Palmyra Atoll, they anchor a magnificent food web connecting herbivore and carnivore, rooted tree and floating plant, animals that fly and those that swim—a single bird joining those that dwell on land with those living in the sea into a life-giving whole.

Douglas McCauley and his colleagues have described how one seabird enriches a food web, making the difference between scarcity and abundance. Green and colleagues have only begun to describe how shorebirds, moving energy across vast distances, enrich life at the sea edge. Who knows what surprise and wonder this story, unfolding along the shores of entire continents, may hold.

Does losing one more bird matter? It's no longer a matter of losing just one bird, or even losing a few. Over the course of Earth's history, animals come and go, the average tenure of an individual species about 1 million years. At that rate, and with a little over 10,000 species of birds living on the Earth today, one would disappear about every 100 years. A person could live his or her whole life without one bird becoming extinct. Yet the pace of loss is picking up. In my own lifetime, at least 19 birds have already gone. One in eight is in danger of extinction: 1,373 species. The IUCN considers another 960 in the category of "near threatened," bringing the number of birds we need to worry about up to one in five.

Darwin, standing amid ancient fossils on the plains of Patagonia, seeing their resemblance to living animals, left us the knowledge that the lives of all species are intertwined, and all who dwell on Earth are kin. Evolution leaves us that understanding, even when some of those connections are visible only when we look back. Fish, for example, would seem to have little in common with humans, yet more than 300 million years ago, one walked ashore, setting a course that ultimately led to our own evolution. Losing one bird out of five has to matter, even when we can't anticipate or articulate how.

In the worst of Earth's five great mass extinctions—the Permian extinction, 250 million years ago—rivers of lava flooded Siberia, setting fire to vast underground coal seams and filling atmosphere and ocean with carbon dioxide. It was a cataclysm that extinguished 96 percent of marine species, but it spooled out over 60,000 years. If we'd been present, would we have recognized the implications of what we watched? Would we have noticed how a greenhouse atmosphere and carbon dioxide–laced ocean became injurious to life? The circumstances were dire, but as we feel the passage of time—day to day, month to month, year to year—we may not see the gradual accumulation of loss or feel the urgency as we, and our children, accept the still-beautiful, greatly diminished world we come into, knowing no other, and not realizing or asking what that world was or could be. Anthony Barnosky and his colleagues at the University of California project that if the current rate of extinction continues unabated, we will have unleashed Earth's sixth mass extinction in just a few centuries. Biologist Stuart Pimm puts it more starkly: "Humanity's impact has reduced species' lifetimes from a metaphorical hour to a minute and it may soon be a matter of seconds."

Losing the knot is not a matter of losing one bird; taking away the homes of the knot, we may lose, sooner or later, other birds living along the sea edge. Jeopardizing the lives of knots jeopardizes hundreds of thousands of shorebirds. Losing, and even beginning to lose, so many birds matters, even we if can't feel how. We are living without the large numbers of whales that once inhabited the sea, but as it turns out, they recycled carbon to the bottom of the ocean. They could be helping us now. In Gloucester, where I live, our lives are impoverished, literally, as a community that once made its living from the sea took too many fish and now is selling its fishing boats and seeking its living by other means. Cod are not extinct, but the rhythm and fabric of life in the sea were altered and weakened, perhaps irrevocably, when fish, once so numerous they could be scooped from the water in buckets, are now so few. Coral reefs have not disappeared, but their vitality is lessened long before the last sea turtle is taken or the last shark speared. Large, vibrant populations of wildlife cease to thrive long before they become extinct.

Already, long before they are in real danger of disappearing, the number of individual birds has dropped, perhaps by as much as 25 percent—millions of birds gone, with millions more to follow. The ecological consequences on Palmyra Atoll from shunting a bird away from its home were profound. For the tropics, scientists are beginning to articulate the ecological and evolutionary consequences of living in a world where large animals are disappearing. They are looking at how the resiliency of food webs and ecosystems weakens when birds that disperse seeds, pollinate plants, or scavenge fish are removed. There is much work still to do on seabirds and shorebirds.

Each time we quantify the value, not necessarily of particular birds or animals but of their homes, the price goes up. When Robert Costanza first assessed the financial worth of Earth's ecosystems, he found that each year, the Earth provides $49 trillion (in today's dollars) worth of services, such as recycling carbon and purifying water. His latest assessment puts that worth at $143 trillion per year. Coral reefs are 42 times more valuable now, not because estuaries and reefs are doing more than they did before, but because we better understand them. Scientists at Stanford, carrying out the same kind of analysis, finding that beaches and salt marshes are worth millions of dollars in storm surge protection, argue to conserve those that remain. These arguments, making a strong case to protect the sea-edge home of shorebirds, will help protect the birds themselves.

Must every bird prove its financial worth? Must every bird serve us? Eiders could probably withstand a cost-benefit analysis. In Iceland alone, their down is worth $40 million, but killing off the passenger pigeon, we couldn't predict its legacy of Lyme disease, and if we'd had to put a price on horseshoe crabs before realizing the astonishing clotting properties of their blood, we would have extirpated them all. If we'd done an economic analysis before DDT brought our national bird to the brink of extinction, what might have been the bottom line for the bald eagle? We may never commodify red knots and other shorebirds. Red knots resize their muscles, hearts, lungs, and gizzards on each long-distance flight, transforming their small bodies again and again from powerhouse long-distance fliers to some of the animal kingdom's most rapid and efficient energy consumers. Perhaps one day scientists will apply knots' physiological flexibility and versatility to a pressing human medical problem, but then again, they may not. We may never fully understand how shorebirds strengthen coastal food webs and how they enrich life at the edge of the sea.

I love walking the beach, finding joy in the summer as piping plover chicks, with their watchful parents nearby, take their first dashes across the sand, and terns wheel overhead or plunge into the water to snare fish for their young. I love walking the beach in spring and autumn, finding comfort in the regular return, season after season and year after year, of dowitchers, godwits, knots, and dunlin on their way to and from distant shores. I love walking the edge of the sea in winter, where I find sanderlings hunkered in the sand, or purple sandpipers in the lee of the breakwater, taking shelter from the wind. What is the financial value of that which nurtures the human spirit? And what kind of uneasy moral terrain do we inhabit when, on the basis of financial expediency, we choose which species will live and which will die?

McCauley writes that saving nature when it can make a profit can help promote conservation, but using economic benefit as the sole basis for deciding whether to protect wetlands, prairies, or beaches is short-sighted and "selling out on nature." The substantial difference between Costanza's first and second calculations, based on our increasing knowledge, suggests that economics is beginning to catch up to what the heart already knows—that the full "value" of nature may never be ascertained, and that a balance sheet, no matter how sophisticated, may never fully encompass or measure the worth of an Earth that gives us life.

If the sixth extinction comes to pass, the lives of our grandchildren and their children may become unbearable and perhaps impossible.

Although the curtain may be rising on this mass extinction, it hasn't happened yet. Biologists who found that we, living as we are living, could bring it about in only a few hundred years, also wrote that "the recent loss of species is dramatic and serious but does not yet qualify as a mass extinction." Much can still be saved, they say, but the challenges are daunting. Red knots speak to us of distant realms, uniting us along a line that stretches along the entire edge of continents. Their long flights, through an immensity of sky that reaches from one end of the Earth to the other, embody our own longings and dreams. In the resilience of shorebirds—in a flock of knots lifting into the evening sky in Bahía Lomas, in a lone whimbrel flying through a hurricane—I find hope and faith that we can face even our most difficult challenges and that a healthy Earth supporting a multitude of species is still possible.

Eleven

THE LONGEST DAY

The Arctic

Finally, I am on my way to the Arctic, where knots will lay their eggs and where a new generation of birds will begin their migration. My first stop is Ottawa, where Environment Canada requires me to spend a day at a shooting range learning to handle a 12-gauge shotgun. At first, it goes badly. The gun has a lot of kick. I am told I will need to keep practicing when conditions are less favorable, that is, when the wind is blowing, when my hands are cold, when I'm not wearing protective glasses and sound-blocking headphones.

After a three-hour flight from Iqaluit, Baffin Island, west across the Foxe Channel, we approach our destination, East Bay, Southampton Island, at the head of Hudson Bay. The landing strip is a piece of gravel ridge just inland from a sea still thick with ice. The bush pilots circle once, coming in low to assess the conditions, then circle again, a dry run for the shortest runway I've ever seen. The third time, we touch down, the Twin Otter's thick tundra tires taking the brunt of stopping quickly in what is essentially a pile of gravel. From the air, the camp looks pretty small—two tiny cabins and two tents in the middle of a bleak, snowy nowhere.

Southampton Island is part of Nunavut, ᓄᓇᕗᑦ, Our Land. Nunavut is the largest and most northern of the Canadian territories and the least populated. No roads connect the 32 small communities dotting Hudson

Bay and the islands of the Canadian archipelago. Southampton Island consists of some 21,000 square miles of rock, river, wetlands, and ponds. People have inhabited the Arctic for thousands of years, but how long they've lived on Southampton isn't well known. European explorers saw or heard signs of possible habitation—rising smoke, footprints, and shouting. Captain George Comer, who made 10 trips to Hudson Bay hunting bowheads, met the island's residents.

Sailing along the island's southern coast in 1896, he saw men and children following him along the shore and, "particularly anxious to make certain inquiries regarding the whaling prospects," which were growing grimmer as whalers depleted the whales, he landed. The Inuit from farther north lived in snow houses and skin tents. Comer found the Sadlermiut on Southampton Island living in huts built of limestone and sod, with roofs framed by whale jaws and windows of translucent seal intestine. The limestone came from a time millions of years ago, when Southampton Island rested on the equator. Comer named Coral Harbor, the island's only settlement, for fossils he found while taking soundings.

In 1899, the Scots established a whaling station on Southampton. In 1902, a Scottish whaler put ashore crew and supplies and offloaded whale oil, ivory, and furs. The sailors also brought with them a severe dysentery that killed all the Sadlermiut except for one woman and her children, who left the island. Today the village is home to 950 Inuit. The summer population of knots could be higher.

Grant Gilchrist, Arctic seabird research scientist with Environment Canada, has generously invited me to join this remote field camp in East Bay, where scientists monitor shorebird nesting habits. Begun by Gilchrist 20 years ago, it's now the oldest continuously operating shorebird field camp in the Canadian Arctic. Meagan McCloskey and Naomi Man in 't Veld from the field team, who'd flown in the day before to set up camp, greet the rest of us—Kara Anne Ward, Alannah Kataluk-Primeau, Gilchrist, the team's leader, and me. Four miles out in the bay, another camp sits on a tiny island, also managed by Gilchrist, where scientists are studying eiders, gulls, and snow buntings. Until a few days ago, heavy storms had blanketed Southampton in fog, rain, and snow, delaying flights to the island camp by two weeks. There are no lights on this tiny runway, and no control tower. When cloud cover is so low it can't be distinguished from snow, pilots don't fly.

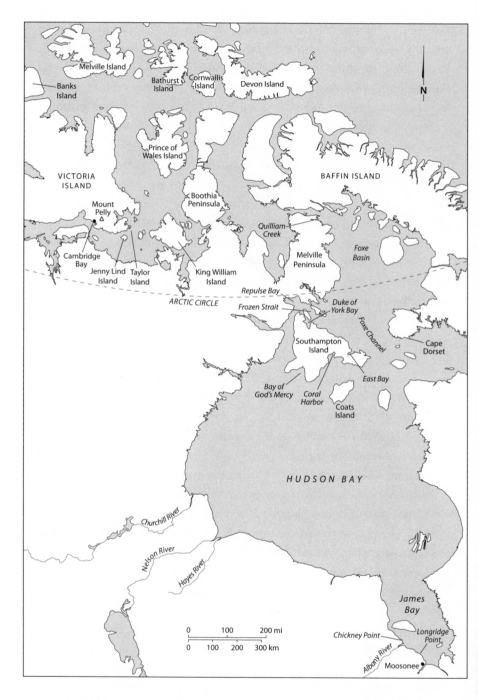

Possible red knot nesting grounds in the Arctic (map by Bill Nelson).

Crew at the island camp had augured the surrounding sea ice a few days earlier. Three and a half feet thick, it could hold a DC-3 outfitted with landing skis and carrying gear for both camps: food for eight weeks; jugs of propane, kerosene, and oil for the heater, stove, snowmobile, and ATV; field computers, radios; guns and ammunition; heavy down sleeping bags; a new tent; toolboxes; emergency survival bags; survey equipment. The plane arrived during a brief interlude of sunshine. Josiah Nakoolak came over from Coral Harbor, 45 miles as the knot flies and three hours across the tundra by snowmobile, to meet the team. Instrumental to the safety and smooth operation of the camps, and an integral member of the team, Nakoolak understands ice, weather, bears, and firearms. He has worked with Gilchrist and his crews since 1996, designing and building cabins, working out ways to trap birds on the tundra, taking the team safely across snow and ice. With McCloskey and Man in 't Veld's assistance, Nakoolak ferried gear to the mainland on his ᖃᒧᑏᒃ, komatik, an Inuit sled with wooden runners. Inuit once shod komatik runners with whalebone, rubbing the bone with wet bear skin to glaze it with ice, ensuring a smooth glide over rough terrain. When the supplies are unpacked, Gilchrist will lend the necessary slipperiness to this komatik with Teflon. Nakoolak was one of the last Inuit on Southampton to pull his komatik with a dog team, but now he too has switched to a snowmobile.

By the time Ward, Kataluk-Primeau, Gilchrist, and I arrive, Nakoolak, McCloskey, and Man in 't Veld have already seen two polar bears. One ran alongside the komatik as they crossed the ice, and another they found dead when they arrived at the ridge. Its hair was beautiful, Ward remembers, like "slivers of glass," but the animal itself was "wasted and thin." Nakoolak thought it had died of starvation. One of the Arctic's largest polar bear populations—2,600 bears—lives in the Foxe Basin. The bears eat ringed seal, patiently waiting by holes in the ice where they will grab seals surfacing for air. Spring is peak feeding time in the Foxe Basin, and bears come consistently to East Bay to hunt baby ringed seals. When the ice melts, they come ashore, many to Southampton Island.

This team is tough and strong. McCloskey served in the Navy Reserves. She is cool headed, quick thinking, smart, and practical—and an accurate shot. In an emergency, I'm not sure I could depend on myself, but I know we can depend on her. At the end of the season, she'll ride 1,600 miles through the western United States on her bicycle. She's worked in East Bay before. So has Ward, who holds a master's degree in biology and will be going to medical school in the fall. Somehow, she's memorized all the complex details of East Bay's research protocols. Her imitations of the courtship songs of shorebirds perfectly reproduce the real calls. Man in 't Veld, a former science teacher who will be obtaining a master's in social work, has field experience studying lizards in the Bahamas and ducks in the MacKenzie Delta. She has a ready smile and an uncanny ability to quietly size up and ameliorate the stresses of this difficult environment. She has overstepped the rules about how much "stuff" we can bring in. For this, we will be grateful.

Kataluk-Primeau, like me, hasn't worked in a field camp before, but she's a quick study, and coming from Pond Inlet on northern Baffin Island, she is attuned to the subtleties of this landscape in ways the rest of us will never be. In the fall, she'll attend Nunavut Arctic College in Iqaluit, where she has enrolled in its program in environment technology. Gilchrist has put together a remarkable team, each member contributing uniquely and critically, and all with uncommon generosity of spirit. He cultivates his teams carefully and leads by example. When he leaves the mainland to join the island crew, the camp runs as if he were still there.

Spring, summer, and autumn are compressed in the short field season of East Bay, a low Arctic location with a high Arctic climate. In the first few days we finish readying the camp: setting up the kitchen tent and hooking in the stove; filling the blue barrels with snow, our water supply; burying a metal box filled with frozen meat; organizing crates of cabbages, onions, apples, and carrots, flats of sauces, packages of noodles, pancake mixes and cereal; and piling up large rocks to anchor the tents during storms. We practice loading and unloading our shotguns and firing rubber bullets and lead slugs at cans set out in the snow. We take the first of many long walks both along the ridge and out to the sea, counting birds arriving as the snow melts.

At home, running three to four miles in my hilly neighborhood and walking two miles in knee-deep water along the beach helped prepare me for these long transects. Still, I hadn't fully anticipated the challenges of walking long distances in bulky layers of clothing, simultaneously managing GPS, radio, binoculars, water bottle, field notebook, and gun, without dropping any in the snow. I hadn't learned how to watch for birds in the sky without tripping on the slippery and uneven ground.

Between the camp ridge and the sea, almost a mile away, the tundra is dappled with ponds, the largest and deepest unceremoniously named Inconvenience Lake. It will, later in the season, when the ice and snow melt, require wide berth, but now we make the crossing. We sink up to our thighs in wet snow and are forced to crawl. The melting snow, though difficult to traverse, carries the promise of spring. Hundreds of geese fly overhead, and Man in 't Veld counts them, easily distinguishing Ross's geese from slightly larger snow geese with their somewhat larger bills, flatter heads, and longer necks, distinctions that come to me only later.

Shorebirds are arriving. A white-rumped sandpiper lands on a patch of exposed tundra, running with one wing straight up, perhaps staking out a territory, perhaps beginning a courtship display. As we approach the shore, ruddy turnstones dart among the rocks. In the days that follow, the lonely tundra fills with the sound of courting birds. In the clear and quiet evenings, we hear them calling across a space that, until they arrived, seemed vast and empty. We hear dunlins, whose calls sound a little like bombs dropping in the distance, the *pee-oo-wee* of black-bellied plovers, the trilling insect sounds of white-rumps, and the soft, muted calls of red knots. Within the first days, we hear or see 12 knots.

One knot is flagged. Kataluk-Primeau and I have been out in the slushy ice accompanying Nakoolak as he gathered goose eggs from moss-lined nests and checked out a seal sitting out on the ice by its breathing hole. I return to find that Ward has spotted, by a few tiny pools of meltwater, six or seven knots chasing each other across the snow, including one with a green flag, 4KL. 4KL's green flag means the bird was banded in Delaware Bay, so I put in a quick satellite call to Larry Niles in New Jersey, who tells me 4KL was last seen in Delaware on May 19. This bird probably arrived in East Bay while we were grounded in Iqaluit. It brought considerably less luggage.

In a few days, we'll hear running water as melting snow turns into streams, but now snow still covers 95 percent of the tundra. It is so quiet I can hear it crunch as the knots pick their way through. Perhaps they are hunting for crane or caddis flies not yet hatched. In these bright evenings, before the nesting season truly begins, these knots that had turned into flying machines are undergoing yet another metamorphosis. Shivering to stay warm in this cold, delayed spring, when there's not yet enough food, and rebuilding their gizzards, hearts, and livers to sizes more suited for successful breeding, they are burning their extra fat and extra large flight muscles.

In the late afternoon light, which lasts until morning, king eiders, a bird I'd previously seen only from a distance, come in to the larger ponds, their magnificently colored orange and yellow beaks in full view. Pink and purple blossoms of saxifrage appear on the ridge, lying low in the gravel for warmth. Knots sing.

During the nineteenth century, when the British admiralty sent ships across the Atlantic seeking the fabled Northwest Passage, what impeded Arctic exploration enhanced the study of birds. Ships immobilized in the ice for nine and ten months at a time afforded their naturalists the opportunity to track birds arriving in the spring. Sir William Parry never found clear passage through the Arctic, but on the first attempt (1819–20), his astronomer and ornithologist Edward Sabine reported that the knot "breeds in great abundance on the North Georgian Islands." Today they are called the Parry Islands and include Bathurst, Cornwallis, and Melville Island, where the expedition wintered, staving off scurvy with mustard and cress that Parry grew himself.

Sir John Richardson reports from Parry's second expedition (1821–23) a knot "killed in the Duke of York's Bay" along northern Southampton Island. (Parry, landing there on the duke's birthday, named the bay after him.) In this brutal environment, evoked by the names of nearby waters— Frozen Strait, Repulse Bay, Bay of God's Mercy—Parry finds knots in Repulse Bay, and George Francis Lyon, captain of Parry's second ship, HMS *Hecla*, discovers a knot nest near Quilliam Creek on the Melville Peninsula, across the Frozen Strait. "They lay four eggs on a tuft of withered grass, without being at the pains of forming any nest."

Such was the cachet of the red knot, a race to find its eggs ensued. Two British naturalists from the 1875–76 Nares expedition came close: after they'd been iced in on Ellesmere Island for 11 months, Henry Wemyss Feilden found a knot and three nestlings, and Henry Chichester Hart, wintering nearby, found nests but no eggs. U.S. Army lieutenant Adolphus Greely's 1881–84 expedition to the far reaches of the Arctic ended in disaster—many of his crew, stranded on Ellesmere Island for three years,

died of exposure and starvation, and the survivors were accused of canni-
balism. Before calamity struck, he did find an egg. He was certain 20 pairs
of knots nested nearby, but "We never obtained the nest. I not only spent
hours in watching a nesting-bird, but had several of my most patient
hunters occupied on similar duty, with no success." On June 9, 1883, his crew
shot a knot, opened the bird, and found, nicked by the bullet, "a completely-
formed hard-shelled egg ready to be laid." One journal heralded Greely's
find as a first—"The Eggs of the Knot (*Tringa canutus*) Found at Last!"—
while another gently offered a correction, reminding readers of Captain
Lyon's observations 60 years earlier.

Richard Vaughan, in his lovely and beautifully sourced book *In Search
of Arctic Birds*, writes, "Of all the Arctic breeding waders whose eggs were
sought and prized by collectors, the Knot took pride of place," so much so
that Americans would continue to claim they'd found the egg first, behavior
that Vaughan finds "hard to explain," given that a 1900–1903 Russian expedition to the Arctic shot at least 14 adult knots and brought back seven sets
of eggs and 26 young birds. The finding was published in English in 1904,
but only a few years later, in June 1909, Admiral Peary's crew, returning
from the search for the North Pole, found knot eggs, with Peary's assistant,
Donald Baxter MacMillan, writing that "the eggs had never been found
previously." They found two sets of eggs in the hills, one and two miles
inland.

In 1916, crew on yet another expedition, searching for what turned out
to be a mountainous mirage and led by the same MacMillan, also claimed
to be the first to find knot eggs. Ship surgeon Harrison Hunt wrote that
knot eggs "had never been found, although ornithologists had searched . . .
for many years, all over northern Asia and America." Botanist W. Elmer
Ekblaw wrote enthusiastically, if not accurately, that "to ornithologists and
bird lovers the world over the most important result obtained by the recent
Crocker Land Expedition . . . was undoubtedly the discovery of the nest and
eggs of the knot. . . . Few eggs have been so eagerly sought . . . yet until this
latest American expedition, the knots had foiled all explorers."

Captain Lyon's knot eggs came from the difficult-to-reach central
Arctic, whose waters were, and still are, lightly traveled by humans. Decades
would pass between recorded sightings there. In July 1853, Robert Anderson,

surgeon on HMS *Endeavor*, shot a knot near Cambridge Bay, Victoria Island. In 1919, Captain Joseph Bernard, frozen in on Taylor Island, adjacent to Victoria Island, collected male birds with downy chicks, juveniles, and eggs. Ornithologist David Parmelee, visiting Victoria and the adjacent islands in 1960 and 1962, heard their mating calls, which he described as *poor-me, poor-me*, from a 600-foot-high esker on Mount Pelly, a full nine miles inland from the sea, near Cambridge Bay. Mount Pelly, he wrote, is "most severe, the summit having been swept almost continuously by the polar winds."

Knots seem to like harsh, wind-scoured places like Mount Pelly or nearby Jenny Lind Island, where Parmelee saw a dozen pairs. He witnessed courtship displays of males, singing and hovering with rapidly vibrating wings, 150 to 300 feet above the island's stony slopes. Far inland, in a low, sandy area of pond and sedge, he saw a knot puffing its feathers and feigning injury. "No doubt this individual had eggs close by, but we failed to find them." He returned to Jenny Lind four years later, heard even more knots, and found two broods of baby birds. In August, as the birds gathered to head south, he found 50 juveniles on the beach—for the Arctic, a veritable crowd.

Scientists still search for knot nests, but as the population dropped, difficulty finding them continued. Larry Niles and Amanda Dey began searching the Arctic for knot nests in 1999 as part of a team that now includes Paul Smith, a research scientist with Environment Canada; Rick Lathrop of the Center for Remote Sensing and Spatial Analysis at Rutgers University; and Mark Peck from Canada's Royal Ontario Museum. Over the years, the team has attached radio transmitters to more than 250 knots and flown hundreds of miles searching for signals. The dry ridge habitat where knots could nest ranges over 77,000 square miles. Over the years, they were lucky to relocate 45 birds. Analyzing data from radio signals, satellite imagery, ground surveys, and even eBird sightings, they continue to refine their map of where, within this vast place, knots are likely to breed.

Since 2002, Jennie Rausch, biologist with Environment Canada, has made seven survey trips to the central Arctic, to islands above the Arctic Circle—Victoria, Jenny Lind, King William, Prince of Wales, Bathurst, Devon, Cornwallis, and Banks Islands—to find shorebirds. She describes a "tale of red knot disappointment," finding in all this time only 10 knots, and no nests. Data loggers may find the few knots still there. In order for engineer Ron Porter to calculate a bird's latitude from a geolocator, he needs to know the time of dawn and dusk, measurements geolocators couldn't record in the Arctic summer's endless day. A new, more sensitive data logger now registers dimming light even in the far north. In 2012, knots began carrying them. One bred on Victoria Island, just west of Cambridge Bay. Perhaps it nested on Mount Pelly, where Parmelee heard courting knots more than 50 years ago.

A satellite tracker would ease the search for nests considerably. Birds carrying satellite trackers, continuously transmitting their location in real time, don't require recapture. However, they must be light and fit comfortably yet securely on small shorebirds whose weight fluctuates dramatically. A unit weighing less than a penny, perhaps available in 2015, may at long last allow humans to follow knots into still impenetrable and perhaps surprising nooks of the Arctic.

Of the 60 birds with geolocators Niles has recaptured so far, at least one nested on Southampton. Southampton, along with King William Island, had the greatest concentration of radio signals. Niles and Dey made the long trip to Southampton eight times, combing remote tundra inland from the Bay of God's Mercy. Smith, no slouch at nest finding himself, says that Niles, Dey, and those working with them have found more nests from knots migrating along eastern North and South America than anyone else. In their best year, 2000, they found 13. In the years that followed, as the population crashed, they found fewer and fewer.

It was early in the twentieth century that Southampton Island, somewhat serendipitously, came under the gaze of ornithologists. Until then, the written record of the island's ornithological landscape, limited to ship logs of whalers and explorers, was scant. George Miksch Sutton, born in Bethany, Nebraska, in 1898, the son of a pianist and a minister, began drawing birds when he was 10. He followed them when he could, once inching his way into an old, rotten log. He became stuck inside with a vomiting turkey vulture, its egg, and a nestling. He went on to become an exceptional field ornithologist and fine artist, traveling to Labrador and Hudson Bay, falling in love with the "clean-edged beauty of the Arctic." Regretting that sea ice relegated his visits to midsummer, too late to witness the birds courting or building nests, he desperately wanted to be iced in for the winter. Then he could greet the spring.

A few years later, on a steamer returning to Montreal, he met fellow passenger Sam G. Ford, factor of the Hudson's Bay Company's new post on Southampton Island. Ford invited Sutton to the post in Coral Harbor, and in August 1929, two months before the stock market crash plunged the United States into the Great Depression, Sutton arrived on the Hudson's Bay Company's supply boat *Nascopie*. "Many an explorer," he writes,

> eager to add a little to what we call the sum of human knowl-edge, or dreaming of wealth and fame, has sailed the polar seas. Many a scientist has fought his way across tundra and through ice-fields hoping to conquer and dispel the apparent inscruta-bility of the frozen ocean, the ice-covered wastes, and the aurora-curtained sky. Many a bird-lover, ashamed that we should know so little about the nesting-habits of some of our common migrant species, has made his way to the inhospitable Barren Grounds, or to the remote Arctic islands, confident of finding there in their summer homes creatures whose domain is the wide world itself. . . . All these explorers, scientists, and bird-lovers acknowledge the charm of the Far North; but none can precisely explain it. One finds one's self while in the Arctic wondering why one should leave a pleasant home for such a cold and savage land, and how, indeed, any race of beings, be they ever so rugged, can continue to remain there. One leaves the Arctic wondering whether one can again resume the noisy intricacies of the over-organized civilization, to which one must return. Safe home from the northern seas, one recoils a little at the thought of winter-long isolation, of frozen face and hands, of insufficiency of food; but one finds one's self dreaming, nevertheless, of returning some day to the magnificent unques-tioning, impersonal friendliness of the tundra.

He lived there an entire year, departing with *Nascopie* the following summer. As he worked in the cold rooms of the trading post, ice crystals clogged the tips of his paintbrushes. On alternate Saturday nights throughout the long, dark winter, if the radio was working, he tuned in for personal messages, including songs, from friends, family, and colleagues, broadcast from KDKA in Pittsburgh after the day's scheduled programming. Subsisting on a high-cholesterol diet of birds' eggs, caribou meat, seal and whale blubber (raw and cooked), he traveled across the island by dogsled, including a trip to East Bay with an Inuit man, Amaulik Audlanat, an "expert mechanic, boatman, huntsman, dog-team driver, and igloo-builder." Sutton owed the success of his expedition to Audlanat, much as Gilchrist has Nakoolak to thank for his.

Sutton wrote more than 250 pages on the habits of the many birds he observed on Southampton, a treatise that would become his Ph.D. dissertation. He found few knots. He never saw their spring arrival, never found their light-green, spotted eggs or young, although he suspected they might breed near the Duke of York Bay. In the fall, a few flocks gathered along the coast, keeping to themselves, "very quiet and dignified in behavior, especially when compared with the turnstones which flashed and rattled along the beaches everywhere."

Other biologists followed Sutton to Southampton, corroborating the idea that knots bred there. Biologists Ken Abraham and Dave Ankney, camping on East Bay in the summers of 1979 and 1980, catalogued 41 species of birds, including, on a visit to a small island, what would turn out to be the largest colony of nesting common eiders in the Canadian Arctic, between 3,800 and 5,900 nests. The biologists were certain knots bred along East Bay, noting "courtship flights, chases, and vocalizations in four separate locations," finding a "broody" female about to lay her eggs, three and a half miles inland and, on the camp ridge, "an adult with one flightless young." Smith has seen as many as 30–40 knots along the shore, gathering for the journey south. As a place to watch shorebirds, and maybe knots, in their summer quarters, East Bay seemed promising.

Inuit who worked with Sutton called June ᐸᓗ, Munniliut, Egg Month. We are eager to find eggs, but downpours and blizzards stymie our search. Rain and snow blow sideways so forcefully that if I try to fall backward, I am held up by wind. It nearly carries away the outhouse roof. Forced inside, we clean our guns, looking for gravel trapped in the barrels. We huddle in the kitchen tent, dressed in long underwear, sweaters, fleeces, and jackets, learning Inuktitut from Kataluk-Primeau, sharing Nakoolak's bannock, a hearty Inuit round bread, and rationing Man in 't Veld's French pressed coffee and special ginger chocolate.

We pride ourselves on culinary creativity: seafood stews made with tins of smoked oysters and crab, curried stir-fries, and from time to time goose, shot and offered to us by Kataluk-Primeau and Nakoolak. They boil one in a fortifying broth with dumplings. They offer us another, which Gilchrist roasts, using a recipe from a 1974 *Joy of Cooking*. Kataluk-Primeau gives us thinly sliced pieces of goose gizzard, raw. It's lined with willow grass. Quotas determine how many caribou and polar bears residents of Southampton may kill each year, but snow geese are overrunning the island (and Inuit can hunt them freely).

In our cabin, wet clothes hang from every rafter. My waders smell of kerosene. The crew members knit mittens, hats, and scarves for almost everyone they know. We crawl into our sleeping bags, the only place where it's truly warm, study bird books, watch Man in 't Veld's movies and, in desperation, read aloud from *Fifty Shades of Grey*. At night, if the wind lets up, we listen to the desolate cries of red-throated loons. We don't know exactly how cold it is, or how wet. During the winter, bears and foxes reconfigured the weather station.

We're required to carry guns everywhere, including the outhouse. Nakoolak has a second sense about bears. Often, I'll find him watching a speck of white tinged with yellow long before it materializes into a large, fast-moving bear. Narrow reinforced doors on the cabin ensure we'll wake up if a bear attempts entry. I hope not to experience this. If bears are heading our way from the island, the eider camp radios a warning. Every day we check in with the team there, reviewing the work and the weather, arranging to share tools or supplies. One day they send over homemade baguettes, still warm. We live in comfort. While I am here, I am given a sleeping bag with many layers of down on both top and bottom. No matter how cold it is, how damp, or how bitter the wind, I am always warm at night.

Tracking shorebirds on the mainland is a world apart from monitoring eiders on the island. Gilchrist began his studies investigating Inuit concerns about declining numbers of eiders. In 1995, Nakoolak and he made the long trip out to the island in East Bay, much as Abraham and Ankney had more than 20 years earlier. After setting up their canvas tent, two-burner Coleman stove, and VHF radio, they sat on the nondescript low-lying rock and waited. Eiders came in, circled the island, and left. They too were waiting.

Eventually six came in, then 10, then 12, then hundreds, then hundreds of hundreds. So fat they couldn't land on the hard ground, they skidded into icy ponds newly forming in the spring melt. Gilchrist and his team, monitoring and banding ducks, determined that too many eider were being killed almost a thousand miles away in west Greenland, where the birds winter. Geenlanders began managing the hunt and shortening the season. The breeding population in East Bay soon tripled.

The eiders that will soon wing overhead are highly prized for their meat, eggs, and down, which Inuit women make into warm parkas, hats, and mittens. Inuit whom Sutton knew called eiders by at least five different names. The knot is less distinguished. Sutton says the Inuit had no specific name for red knots, considering them, along with semipalmated and white-rumped sandpipers, simply shorebirds. In Iqaluit, I met Inuit who'd flown in from all over the Arctic to consecrate St. Jude's Cathedral, newly rebuilt after a fire, and to celebrate the long-awaited Inuktitut translation of the Old Testament. Over lunch, they told me they knew the knot, and that knots and their eggs were too small to bother hunting. They, like Nakoolak, referred to small sandpipers, including knots, as ᕈᖅᓯᓇᐊᖅ, *sijjariaq*, birds of the beach.

When the weather clears, we begin our nest search. We pack lunches of leftover dinner, tuna fish, or peanut butter, along with tortillas, and when they run out, pilot biscuits, long an Arctic staple. In old photographs of Iqaluit schoolchildren, pilot biscuit boxes, recycled to hold school supplies, sit on every desk. The hard, dense cracker, made from salt, water, and flour, never breaks in the crush of our equipment or when I stumble and fall on the tundra. Rather dry to begin with, it never gets stale. It lacks flavor, but is virtually indestructible.

Common eiders breed in dense colonies: when Gilchrist's study began, 50 eiders nested in each plot. As the population recovered, the number grew to 250. Shorebirds are more solitary. The mainland study area, 3,000 acres compared to the island's 60, might yield fewer than 10 nests per plot. We hunt for seven or eight hours each day, scrambling over ice floes and wading through ponds. Stepping into a dark pond, I never know whether I'll slip on ice or be sucked into mud. Feeling the pressure on my waders in the deepening water, I am grateful for seams that hold. When the weather permits, we can walk eight or nine miles a day. Sometimes at noon, we'll take a short break, sitting in the warming sun, peeling off our jackets. More often, it's bitter cold. On a tundra scoured by ice and storms, the occasional boulder offers a brief respite from damp winds coming off the ice. We're slathered in sunscreen, even on gray days, but my hands blister.

Even in the cold, I love the peace of East Bay, its wide, open expanse and the quiet, stark beauty of its dry gravel ridges. As the season goes on and the snow melts, I love the sound of running water, singing birds, and the almost overnight appearance of tiny clumps of flowers and myriad hues of green amid the rocks—life persisting in a difficult place. Accompanying Nakoolak out on the ice one day, we stop at a distant ridge, where the only sign of humans passing through is a sun-bleached walrus skull resting atop a small, windowless plywood box where an Inuit hunter had spent the night.

I am in a well-run, well-provisioned camp, but I have no illusions that I have the skills to live here, or even visit this harsh place without substantial support. One evening, Gilchrist and I listen to a radio broadcast from Canada's Polar Continental Shelf Program, an agency providing logistical support to Arctic researchers. Transmitting from Resolute Bay, more than 700 miles north of us, it checks in with field camps spread 1,500 miles across the Arctic, from Alert, Ellesmere Island, to Inuvik on the MacKenzie Delta. In this seemingly empty place, we are anything but lonely.

The tundra between the camp ridge and coast is busy with birds. On bare rocks in the middle of the ponds, herring gulls build nests of dried kelp and the rare twig they take from the tide line. Arctic terns are laying their eggs in shallow hollows, the eggs barely distinguishable from gravel. The terns, graceful and buoyant fliers, reveal the presence of nests by dive-bombing our heads as we approach. They are among Earth's longest long-distance fliers, each year traveling 50,000 miles back and forth between Earth's iciest realms. Gray-hooded Sabine's gulls—named after Edward Sabine, who first spotted them during Parry's search for the Northwest Passage—nest close to the coast, making a racket if we approach.

The nests of shorebirds—ruddy turnstone, semipalmated plover, and white-rumped sandpiper—lie ingeniously hidden in plain sight, a few pieces of lichen in a sheltered spot amid the stones. Unlike the Wilson's plovers in Georgia, whose nests I could find following the birds' tracks across the sand, these birds don't leave tracks. We accidentally stumble upon the nests or, when a bird flushes, approach the nest when it returns. If a bird does flush, there are few signposts in the shattered rock to guide us to the nest. If we're kneeling and we stand up or, if standing, we take only a few steps, perspective and depth perception dramatically shift. Rocks that seemed adjacent are no longer so, and I cannot distinguish markers that had seemed fixed—two lichen-draped rocks here, a hollow there—from thousands of other rocks and hollows. The others magically maintain their sightlines, reading an impenetrable landscape.

Smith joins us after storms have delayed his arrival by several days. The man has limitless energy. He bounds through the tundra in the pouring rain, shouldering his gun barrel down, stock high, to blunt attacks from angry gulls and terns, and easily picks out camouflaged eggs. No ruddy turnstone can hide from him. We even find knots leisurely swimming amid the rocks in a small pond. This sight, unfamiliar to me, is historically known to ornithologists, who've found knots "able to swim with great ease." Here, they do swim with ease, seemingly leisurely and tranquilly, a contrast to the feeding frenzies I've seen in Mispillion Harbor.

Nest tracker by day, electrician by night, Smith, after a full day out in the field, mounts solar panels on the cabin roof and wires in the generator to recharge the GPS, radio, satellite phone, and computer batteries, obviating the need for gasoline. He also installs two lightbulbs, a seeming luxury during the Arctic summer but of great use in the dark, stormy days and weeks ahead. He isn't finding the ruddy turnstone nests he hopes to see. The season appears delayed, a concern when turnstones breeding here and migrating along the Atlantic coast are steadily and substantially declining.

Walking in a plot near the coast, I'm peering ahead for flighty shorebirds flushing in the distance when Smith points out a phalarope sitting at my feet. These beautifully red and gray birds feed in tundra ponds, rapidly spinning in circles, creating a vacuum that sucks their prey—the larvae and eggs of small aquatic insects—up toward the surface. They are rugged, spending nine or 10 months at sea, coming ashore only to breed, and braving severe storms to feed when other birds hunker down. Inuit hunters mount them on kayak prows as protective talismans. Phalaropes are also declining. Smith doesn't find the number he expects by this time. He thinks perhaps they are waiting out the storms at the floe edge.

We walk inland from camp, to wet meadows where dunlins nest in small, perfectly formed cups of grass. Ward finds a stilt sandpiper, possibly the first recorded for Southampton. We walk along high ridges where black-bellied and golden plovers flush 300–600 feet away from us. Once one flies from its nest, we await its return, creeping closer ever so slowly. The mate dashes down the tundra feigning a broken wing. If we even glance at this distraction, perfected to fool predators, we'll lose our orientation to the nest. Kataluk-Primeau patiently watches a black-bellied plover for two hours before the bird leads her to its nest.

Ward, seeming to see in all directions simultaneously, has a knack for picking out clusters of small rocks likely to shelter eggs. She crouches or lies down, and when we're so stiff with cold we can't take much more, the bird, often a white-rumped sandpiper, will walk onto its nest. Occasionally, that's also a distraction. We've seen semipalmated plovers sitting, tail feathers up, as if they were nesting, while the mate sneaks back to the real nest some-where else. Other predator avoidance we can't see. Sandpipers preen with wax excreted from an oil gland at the base of their tails. As incubation begins, the wax composition changes to a more viscous but less volatile oil whose smell foxes may detect less easily.

Red knot nests are elusive. Unlike plovers, knots don't readily leave their nests, sitting tight unless they are in immediate danger of being trodden upon. Ward monitored one knot nest in East Bay in 2009. Her colleague, Darryl Edwards, an evolutionary ecologist who observed nesting knots in East Bay for three years, found it. "We see knots regularly at East Bay but find nests only infrequently," he wrote me. In 2007, midway through July, a little over a mile inland, Edwards saw "a defensive parent. . . . This bird probably had chicks by the way it was behaving." He believes the knot, having nested inland, was en route to the coast. In 2008 the team found one nest, on July 7. A jaeger plunders it two days later. In 2009 the team found two nests, one on June 25 with four eggs. It fails of unknown causes 10 days later. (A jaeger may have taken the eggs but left no evidence.) On July 3 they found the other nest, and on July 12, four chicks. In 2010 they found "a knot with three chicks." We're all looking, but so far with no success.

Our tasks here, if difficult, are few: ensuring our safety, taking care of each other, surveying the landscape. The barrage of information to which I'm accustomed back home—from e-mail, Facebook, newsletters, newspapers, and radio—is absent here. In the stillness that opens it seems as if I can feel the Earth breathing. It seems an almost unimaginable feat of endurance that tiny birds could year after year fly thousands of miles to build nests, incubate eggs, and bring forth a new generation in this harsh, spare landscape at the edges of Earth's northerly lands, at the other "uttermost end of the earth." Summer, if it ever arrives, will be brief. As soon as eggs are laid, they can be eaten. For shorebirds this far north, long, light-filled days, abundant insects, fewer parasites, and less predation offset the substantial metabolic and energetic costs of getting here. Over an increase of just 29 degrees of latitude, between James Bay and northern Ellesmere Island, the rate of nest predation drops 65 percent. Still, the statistics are grim. Smith tells me that in a good year, less than 60 percent of the eggs will hatch. In some years, it may be only 10 percent, and many, many chicks are eaten before they fly. If a nest fails in this short Arctic summer, there's often no time to try again.

Predation abounds. We see Arctic foxes, their white fur whiter than snow, trotting across the tundra carrying goose and eider eggs in their mouths, caching them for later retrieval; herring gulls harassing loons, forcing them into the water, plundering their nests; and hawklike parasitic jaegers gliding low across a ridge, swooping down on adult birds. In the stomach of a parasitic jaeger, Sutton found a phalarope swallowed whole. Red knot parents take turns incubating, behavior that leaves the nest less exposed. Smith, hiding video cameras near nesting shorebirds in East Bay—work often described as his "nanny cam"—found that other things being equal, when both parents share incubation, four times as many nests survive as when only one parent incubates.

If the menu for foxes and jaegers includes small, furry lemmings, shorebirds may fare better. Lemming numbers regularly surge and then fall precipitously, for reasons not fully understood. If this were a lemming year, we'd find these rodents everywhere—out on the tundra, scurrying about the cabin at night, in our boots. We don't see any, another sign, along with the weather, of a potentially hard summer. The nesting success of snow geese on Bylot Island, brant and knots in Siberia, and turnstones in East Bay has been linked to the abundance of lemmings: increasing when lemmings

On the red knot nesting grounds (drawing by Michael DiGiorgio).

are plentiful, falling when they are scarce. Scientists comparing trends in lemming abundance with nineteenth-century hunting records wonder whether a dearth of lemmings may contribute to the decline of knots. Others think the collapse of lemmings may be linked to the warming of the Arctic.

We have a few glorious days. Tiny willows, no more than an inch or two high, bloom. Our nests now have four eggs. Knots are nearby, but we haven't seen a nest. The evening light doesn't fully fade, bathing ridge and ponds in a soft, rosy hue. The fair weather doesn't last. It's July, ᐸᑦᐃ,

Kittuailliut, "mosquito time," but mosquitoes and other insects whose copious numbers will feed baby chicks aren't hatching in this cold. We wake to pounding rain. Kataluk-Primeau teaches us an old Inuit adage that translated means "It can't be helped." We repeat it frequently.

The rain lets up, barely. We can't stand it inside anymore. When the wind lessens, we head out, hoping it will clear. It doesn't. Instead, on a distant plot of tundra, meadow, and pond, McCloskey and I are enveloped by thick fog. The visibility dangerously low, we slowly retrace our steps, backtracking along the circuitous route recorded on our GPS. A faint silhouette appears in the mist only a few feet away—a caribou. Knots sing softly. McCloskey orients herself by a herring gull's nest taking shape in the fog, and then by a tundra street sign, an *inukshuk*— △ₒᵇ↻ᵇ —a stone cairn assembled by the Inuit to aid navigation or to identify a hunting ground or cache. About where she expects, the camp ridge emerges. The mainland field season is almost half over. Summer is passing.

Soon the knot eggs, if there are any, will hatch. On the 1900–1903 Russian Arctic expedition, Hermann Walter observed that "the male was always most careful of the young, whereas the female, when in the vicinity, had the appearance of a disinterested spectator." She doesn't stick around long. After the eggs hatch, the mothers depart on the long journey south, their job completed. Baby knots leave their nests within hours of hatching, able to feed themselves. Their fathers accompany them from dry ridges down into wetlands where mosquitoes will hopefully abound, walking great distances, as much as two miles, with their flightless young. Adult males also depart before their chicks, who find their way to wintering grounds on their own. After the shorebirds and the ducks are gone, belugas will swim through East Bay, making their way through Hudson Strait to their wintering grounds, to a winter growing increasingly warm.

Not long after we arrived in East Bay, sturdy ice that held a heavy DC-3 was softening into slush that slowed, then nearly engulfed the wheels and runners of our snowmobile and komatik. Nakoolak, Kataluk-Primeau, and I spent an unnerving hour far from camp extricating them. Greenhouse gas emissions are warming the Arctic at rates three to four times higher than the rest of the Earth: summer temperatures in Arctic Canada are the warmest in any century in the past 44,000 years. In 1979, scientists began monitoring Arctic sea ice with satellites. Summers between 2007 and 2013 brought record losses. Thicker, older, more stable ice, persisting four or more years, is disappearing. This more resilient ice, which at the end of winter once made up one-quarter of Arctic sea ice, had by 2013 dwindled to a mere 7 percent.

Melting ice triggers a cascade of effects that ripple through Arctic food webs, touching the lives of nesting birds. Each summer, some 30,000 pairs of thick-billed murres nest on the cliffs of Coats Island, just south of Southampton. The black-and-white birds incubating their single eggs look like small penguins. As summer begins, their view from the cliffs is of sea ice. Scientists scrunched in wooden blinds watch murres feeding their young, one fish at a time. In 1980, Arctic cod, cold-water fish living among the ice floes that produce their own antifreeze, dominated the nestling diet. Now, the ice disappears three weeks earlier than then, the cod are moving away, and parents are serving capelin, sand lance, and blennies. It takes two and a half skinny capelin to provide the same energy as one bulkier cod. Nestlings, less well nourished, gain less weight, and leap from their ledges into a life at sea 10 percent lighter. The population, once increasing, may be leveling off.

In another new twist, murres are increasingly besieged by mosquitoes "so thick," says biologist Anthony Gaston, who oversees the work of Coats Island, the birds look like "they are wearing fur boots." The summer of 2011 was particularly severe. When the crew arrived, murres on their nests were succumbing to heat prostration and mosquito attacks. Polar bears made matters worse. Single polar bears had made brief forays into the colony before, but this time as many as four polar bears stalked the camp each day, nimbly scaling the steep cliffs, squeezing onto ledges only 16 inches wide, dislodging or killing eggs and chicks. They killed adults, one every 10 minutes. Adults nesting nearby panicked and fled, their eggs perishing from exposure. Bears took 30 percent of the nests. Continued losses like these could seriously threaten the colony.

Gaston has watched nesting thick-billed murres on Coats Island for over 30 years, measuring weight loss in nestlings as they feed less on Arctic cod and more on capelin and sand lance, and tracking the increasing numbers of polar bears and mosquitoes in the colony. Polar bear incursions, he believes, are transient; as the Arctic warms, the bears will decline.

Thick-billed murres have nested on Coats for 2,000 years, but as the Arctic warms, they too will disappear, razorbills and puffins coming north to replace them. Razorbills are already arriving. Seabirds, he says, will continue to breed in northern Hudson Bay. "It just won't be the Arctic anymore."

In late spring and early summer, before the ice breaks up, polar bears have historically accumulated as much as two-thirds of their energy needs for the entire year feeding on ringed seal, whose plump, freshly weaned pups are almost 50 percent fat. At the beginning of the season, ringed seal often haul themselves out of the water onto sea ice near the East Bay eider colony. Compared to 35 years ago, loss of summer sea ice along the Coats Island murre colony and in East Bay is now costing polar bears two months of seal hunting every year, forcing them ashore earlier.

In East Bay, bear incursions into the eider camp are seven times higher than in the 1980s, with devastating consequences. The colony has been ravaged, first by avian cholera, and then by polar bears. Now, eider reproduction is too low to sustain the population. The crew scares polar bears away from the island's 5,000 nests, screaming and shooting off cracker shells. About seven bears, including a mother and her cub, visit the camp, returning repeatedly to forage on the eider egg buffet. Sated, the bears sometimes fall sleep. Of several hundred nests left at the end of the season, only seven or eight are still intact when the frustrated and exhausted team leaves, a week and a half early. The following year will be worse, with two or sometimes three bears in the camp once a storm pushes all the ice out. Persistent polar bears are a growing danger: in future years, the team will depart before the ice breaks up, leaving video cameras to record the behavior of ducks and bears.

East Bay is by no means an anomaly. During years of record low ice coverage, between 2010 and 2012, polar bears raided eider colonies on 70 islands along southern Baffin Island and northern Québec, resulting in near-total reproductive failure in most cases. On a small island off Cape Dorset, Baffin Island, across the Foxe Basin from East Bay, when bears finished going through 334 active nests, only 24, less than 10 percent, remained. Colonial nesting on small islands, once a successful adaptation to the threat from Arctic foxes, is becoming a liability. Gilchrist is now concerned that eiders, having recovered from overhunting, are again at risk. "If bears continue to come ashore early," he says, "the ducks have no defense." Some East Bay eiders, he has observed, are beginning to adapt. "Ducks left the mainland to escape foxes, but now a few return, nesting closer to town where humans, their former predators, provide safety." Gilchrist and his colleagues are watching firsthand an Arctic changing as the Earth warms and sea ice retreats.

Shorebird nests, dispersed across the tundra, don't seem to interest hungry polar bears. A more immediate threat originates in agricultural fields on the Great Plains, where increased fertilizer use has doubled corn and tripled rice yields, leaving ample spilled grain to feed wintering and migrating geese. Well-nourished, the number of Ross's and lesser snow geese flying through the Mississippi Valley and across the Great Plains had skyrocketed to at least 16 million by 2009, despite relaxed hunting regulations designed to curb the population.

Southampton Island now also hosts 940,000 nesting geese. Hungry polar bears are finding their eggs. In 2004, on the Bay of God's Mercy, a lone polar bear, over the course of two weeks, consumed the entire contents of 400 goose nests. In 2006, the field team at Coats Island watched bears systematically ransacking a goose colony, targeting nests, devouring eggs. In 2010, James Leafloor from Environment Canada, conducting an aerial survey of nesting geese on Southampton and Coats Islands, counted "about 29 polar bears, many of them in snow goose colonies, far inland from the coast." A summer diet of Arctic char and berries has proven no substitute for ringed seals. Polar bears may kill a ringed seal between one and three times a week. It's hard to imagine how tiny goose eggs, no matter how plentiful or rich in protein and fat, can fully substitute for 100-pound ring seals and their blubbery pups, compensate for a seal hunt now reduced by two months each year, or offer a long-term reprieve for animals turned toward extinction as the sea ice retreats. Karyn Rode from the U.S. Geological Survey and her colleagues, analyzing this question from previous studies across the Arctic, are finding that based on data in western Hudson Bay, bird eggs might at best offset only a day or two of time bears now lose foraging at sea.

Geese crop Arctic grasses and sedges, and, "grubbing," can go on to dig out the roots, impoverishing nesting grounds, turning a freshwater meadow into a carpet of bright green moss, or a coastal marsh into a saline mudflat where little grows. Where geese nest in East Bay, communities of lichen and heath are thinning, sedges are disappearing, and meadows drying.

In some coastal plots on East Bay, it's hard to avoid stepping in goose droppings. Walking along the ridges, I see groups of 40, 50, 100, and 200 geese standing in distant patches of muddy, denuded meadow. Smith wonders whether geese are encroaching on the nesting areas of semipalmated sandpipers, whose eastern population he describes as "tanking." In East Bay, he says, the birds are still "thick" in inland marshes unfrequented by geese, but where geese are abundant, semipalmated sandpipers are gone. The team finds only two semipalmated sandpiper nests. For now, knots seem protected by their remote and desolate nesting sites.

Soon, however, their lives may be brushed by Arctic warming. This cold summer on Southampton Island, I don't see intimations of that world—the arrival of competitors, predators, and parasites from the south, or a timing mismatch between the hatching of baby birds and the explosion of mosquitoes—but the melting sea ice heralds the changes to come. By 2080, as the temperature rises a projected five to 11 degrees and the tundra contracts, sedges and grasses will yield to evergreen shrubs and trees. Forests will begin growing on Southampton. Perhaps the knots can move north, into refugia that remain. They have before. They survived one bottleneck as advancing glaciers turned much of their tundra breeding grounds to desert, and then another, as the Earth warmed and the treeline moved north, but their numbers crashed. What will happen in the coming years as the tundra once again recedes is a question whose answer is unfolding.

Gilchrist, Smith, most of the duck camp, and I are leaving. When the Twin Otter arrives, I walk to the edge of the ridge. Looking down, I see an enormous paw print. The ice is breaking up. When the plane takes off, it ascends quickly. I peer from the window as the camp, my whole world for three and a half weeks, shrinks, then slips away. I will miss the company of the generous, kind, talented scientists who welcomed me into their summer home and who, compiling their data year after year, witness and record how human endeavor so very far from here nudges this land and seascape, redesigning it, restructuring the web of life it supports, sending some animals on their way out, perhaps permanently, and inviting others in as, in nuanced and surprising ways, a land cloaked in snow and ice shakes off the cold.

Our summer ends poorly for shorebirds. Gilchrist ranks it as the worst in the program's entire history. On one dispiriting day, at the end of the season, the team called me. Except for a few days of sunshine, the stormy weather, chilling cold, and high wind hadn't really abated. By the end of the season, they had found 56 nests in all, 25 percent fewer than last year, 50 percent fewer than the year before. Of the 56, eggs hatched in only eight, a dismal finish. Birds were departing, Man in 't Veld said, and the tundra had grown eerily quiet.

On their very last day, Kataluk-Primeau, Man in 't Veld, McCloskey, and Ward fanned out across the plots, checking the remaining nests one more time. Ward wrote me that Kataluk-Primeau found the knot first, near the camp on the next ridge inland. It flew toward her, landed, and "did a broken wing display!" She couldn't find the nest; the bird "kept disappearing and reappearing." Man in 't Veld came by later and "again encountered the knot. This time it called a lot and lifted both wings straight above its back." She couldn't find the nest either. Ward tried. "About 600 feet from camp a bird flushed and did a broken wing display." She ignored it, edged toward the nest, and found "almost right away, a nest full of fluffy nestlings. I thought I had located the knot nest!" She looked up, and to her dismay, saw the mother—a white-rumped sandpiper.

Straight ahead of her was the knot. "I had inadvertently stumbled across a white-rump first. I quickly marked the nest and tried to focus on the knot. It called a bit and raised both wings above its back. It then ran quickly south. I was amazed how difficult it was to follow the knot; it blended in very well with the rocks. I lost track of it over a ridge. I sat for a while hoping to spot it again, but unfortunately did (k)not." They all decide to eat quickly, finish packing, and return for a last look, but their plane, taking advantage of a momentary clearing in a day filled with fog, arrives unexpectedly, and they have to leave. In Iqaluit, the team calls Smith, who thinks the knot's behavior suggested fledglings nearby.

After the nesting season, many knots leaving the Arctic gather to feed in James Bay, at the other end of Hudson Bay, before returning south. Perhaps I'll see these birds there.

Twelve

RETURNING SOUTH

James Bay, the Mingan Islands, and the Guianas

A small, rustic Cree hunting camp lies hidden in a forested rise on the western edge of James Bay, one of a number of similar camps dotted along a coast rimmed with marshes, gravel ridges, and broad mudflats exposed by six-foot tides. The nesting season is over, and now at least 1 million shorebirds will fly in to feed in James Bay, including at least 30 percent of knots heading home along the Atlantic flyway. At one camp, Longridge Point, I hope to see them: perhaps a wave of adults or, depending on how the nesting season has gone, juveniles.

The trip to this remote point begins with a long drive from Toronto with Jean Iron, a retired elementary school principal who, along with the rest of the crew, belongs to the crème de la crème of Ontario birders. Her small white Toyota, packed with gear, has had its fair share of time on the birding road as well as run-ins with deer: a friend suctions out the smaller dents with a plunger. We head north, crest a small hill—elevation about 1,000 feet—and dip into the Arctic watershed, where streams and rivers drain north into James and Hudson bays and out into the Labrador Sea. I barely feel the climb or the descent that follows as the landscape begins answering to a distant, Arctic call.

We detour down unpaved roads. I catch a whiff of evergreen. Iron gets out, points her binoculars up into the trees. Boughs of white pine droop, heavy with cones. White and black spruce are also laden with cones. Dry catkins hang from white birch and trembling aspen. From this information, Iron's partner, Ron Pittaway, makes his annual winter finch forecast, telling Ontario birders where they are likely to find red crossbills, white-winged crossbills, grosbeaks, siskins, redpolls, and purple finches. Another van, driven by Christian Friis of the Canadian Wildlife Service, is also heading to James Bay. Friis coordinates summer shorebird monitoring there, a joint project of the Canadian Wildlife Service, the Ontario Ministry of Natural Resources, the Royal Ontario Museum, and Bird Studies Canada. Although others may shy away from weeks of trekking through mosquito-ridden marshes, Friis relishes remote fieldwork. Not surprisingly, his daughter Sabine's first word was *bird*.

Friis's van stops in Powassan, a tiny town of about 1,000. Their detour is to the town's sewage lagoons, where controlled water levels and nitrogen-rich vegetation provide important stopovers for migrating ducks and shore-birds and, according to team members Mike Burrell and his brother Ken—who've been birding since they were children—great viewing from elevated dykes surrounding the ponds. Ontario birders looking for a lagoon can check a special map with 90 sites. In Ontario, Mike Burrell says, "every serious birder visits multiple sewage lagoons many times." They hope to see a red-necked phalarope in Powassan, but don't.

After 10 hours on the road, we arrive in Cochrane, a town half French and half English, birthplace of Tim Horton, former star hockey player on the Toronto Maple Leafs and founder of a large chain of popular coffee and donut shops we'll patronize frequently. The main road, following the Trans-Canada Highway, turns west. We continue north, taking the Ontario Northland Railroad through forests of pyramid-shaped white spruce, skinny black spruce with knobby tops, and small, wispy tamaracks whose slender needles drop in autumn. We travel along big rivers—the Abitibi, the Mattagami, and the Moose—passing two hydroelectric dams. The team looks for sharp-tailed grouse, northern hawk owls, and hawks on the power lines.

Five hours later we arrive at Moosonee, the end of the line. James Bay was the end of the line for Henry Hudson, who sailed into Hudson Bay in 1610, believing a body of water this huge would carry his ship, *Discovery*, through the Northwest Passage. Instead he entered James Bay, a dead end, where the angry crew mutinied, casting his son and him adrift in a shallop, never to be heard from again. Fifty years later, the king of England granted 3 million square miles of Hudson Bay watershed to the Hudson's Bay Company, whose fur traders set up trading posts along the bay. Andrew Graham and Thomas Hutchins, working at the company outpost at York Factory, at the mouth of the Hayes and Nelson rivers, kept extensive journals, recording in 1770 what may be the earliest sighting of a red knot in North America. Today, about 14,000 Cree live in Moosonee, Moose Factory, and villages along the coast.

We have some time: we're in the middle of a vast, undeveloped wetland, and the team goes out to Moosonee's sewage lagoons to look for birds. They spot 15 ring-necked ducks, 10 lesser scaup, and 75 goldeneye. They stop at a stand of trembling aspen, calling, *psshh, psshh, psshh* to attract a possible mixed flock of warblers. Mike Burrell's cell phone rings: it's programmed with different bird calls for each friend and family member. When the helicopter arrives, we split up, heading to three different Cree hunting camps whose owners rent them to the project.

Longridge Point, 35 miles north of Moosonee, sits in a copse of spruce and tamarack near a long gravel ridge that runs along the shore. For two weeks, we will spend our days walking, setting our clocks to the tides, and counting shorebirds. Barbara Charlton, an avid, longtime Ontario birder and one of the province's top listers, has conducted bird surveys for over 30 years. Iron, she, and I take one of our first walks south to Beluga (also called Paskwachi) Point: white pelicans float in the water, Caspian terns stand on rock islands, and ruddy turnstones, Hudsonian godwits, and knots feed on a quickly disappearing mudflat. We'll often see Hudsonian godwits here, many in flocks of at least 200 birds. I'm happy to see them: they fly nonstop from James Bay to South America, only occasionally stopping in the United States. Charlton and Iron count 58 knots. Two are flagged. It begins drizzling, and then it pours. The knots take off, heading south. The next day Beluga Point is full of sandpipers—25 least sandpipers, 1,110 white-rumps, and 179 greater yellowlegs. Many are juveniles, their plumage shining and fresh. They aren't frightened by humans yet; we approach quite close, crouching low in the rocks.

Near the end of the point 70 knots, splashed by the spray, feed in seaweed at the edge of the rocks. Compared to East Bay, the point is crowded.

More than 500 fly overhead, going south. Longridge has long been a favored spot for migrating birds. At the end of July 1942, Clifford Ernst Hope and Terence Michael Shortt from the Royal Ontario Museum canoed 100 miles from Fort Albany south to Moosonee. Over 10 days they witnessed the southern migration of adult shorebirds, including at dusk one day "an enormous flock . . . stretched out for a distance of a mile or more as a dense, long cloud." In 1976, Morrison and Harrington, flying over James Bay, saw their greatest concentration of knots—5,000—on Longridge Point, and in the years that followed, as many as 15,000 in the bay, confirming it as a key stopover. In James Bay, Morrison worked in the air and in the mud. Geologist Peter Martini and he took cores of mud along 800 miles of James Bay coast, and found the food that fuels the birds' long flights south—dense concentrations of tiny clams and mussels, between 2,000 and 9,000 per square yard.

Quiet fills our days. We see few people; once or twice a canoe passes by. A winter road built of snow and ice serves the small communities along the coast; when it thaws, supplies are barged or flown in. To the west are 144,500 square miles of bogs and fens—the Hudson Bay Lowland, the largest wetland in North America and the third largest in the world. It's lined with impermeable clay, the bottom of a once-broader Hudson Bay. Dipping almost imperceptibly to the shore, the flat waterlogged plain barely drains. The Ontario lowland alone contains 21,000 bogs, inland lakes, and ponds. Dying plants pile up in layers of soggy peat between five and 14 feet thick. The muskeg is laced with ponds; sphagnum moss covers the bogs. Stunted black spruce and trembling aspen take root on slightly elevated and less saturated stream embankments.

While Charlton, Iron, and I count birds at Longridge, Friis, the Burrells, and Jeanette Goulet are 62 miles farther north at Chickney Point, north of the Albany River. The site is mired in mud, so much and so deep that Ken Burrell, sick of continually extricating himself and his boots, gives up and goes barefoot. In the evenings, they compensate for these travails, partaking of delicious Mennonite summer sausage brought in by the Burrells and also by draining the syrup from an enormous can of fruit and replacing it with a bottle of Havana Club. At Chickney Point, they can't approach the birds as we do—it's a full day's slog to even reach the bay—but every day, they watch the flyby. They begin counting two hours before high tide, tallying numbers that defy their expectations: close to 1 million during their two weeks and over 100,000 on a single day. The daily highs for individual species reach into the thousands: 28,570 white-rumped sandpipers with lows never below 10,800; highs of 88,130 semipalmated sandpipers; and 19,420 dunlin. Knots are another matter, a mere 391 altogether, with a high of 125 and one juvenile.

From Longridge, unlike Chickney, it's an easy walk to the shore, passing through a wide sedge marsh filled with wildflowers, and on warm, still days, mosquitoes, as many as 5 million per acre. Even the most stalwart among us wear gloves and special lightweight, densely woven jackets and netting that keep out the hordes of blood-thirsty insects. As we walk through the marsh, we sometimes see what I take at first to be dead tree trunks—out of place here, considerably thicker than the slender spruce and tamarack near the cabins, and appearing in slightly different locations each time. They turn out to be mother black bears on their hind legs, looking at us, their cubs' ears just visible above the flowers. They are eating strawberries.

Every year, the shoreline of James Bay grows wider, and the distance from camp to the ridge grows longer as land around Hudson Bay, once pressed beneath a glacier two miles thick, springs back. When the last Ice Age waned 7,500–8,000 years ago, water poured into the depression left by the glacier, drowning parts of Nunavut, including Southampton Island, and Ontario and Manitoba beneath the Tyrrell Sea. As land released from the heavy weight of ice rebounds, the Tyrrell Sea and its remnants—Hudson Bay and James Bay—shrink (James Bay is only 150 feet deep now) and what was once water is becoming shore. Only 30 years ago, a camp south of Longridge at Shegogau Creek once perched at the high-tide line. Today, it's the length of two to three football fields away from shore. Rising seas are not an imminent threat here.

As land edging Hudson Bay emerges from the sea, its geologic past emerges as well. The walk to Beluga Point meanders through a mudflat strewn with rocks and boulders. Tracks of bear and moose, current inhabitants of the landscape, cross the mud, passing through fragments of ancient coral reefs—pieces of honeycomb and horn coral from a time some 445 million years ago when these lands belonged to a warm, tropical sea. One of our team, Andrew Keaveney, will find a fossil of a trilobite here. Trilobites, animals with many pairs of jointed legs, hard carapaces, and antennae, were once ubiquitous in the ocean, swimming, burrowing, and crawling in the water. Horseshoe crabs once shared the sea with these ancient animals. They endure, while the trilobites, some 15,000 species, are now extinct, along with the coral fringing the old seafloor where we now walk. David Rudkin from the Royal Ontario Museum tells me the trilobite fossil may have come down from Southampton Island, carried by glaciers on a long and slow journey over distance a knot will fly in a day.

Keaveney, Ian Sturdee, and Josh Vandermeulen hike the long, muddy, 14 miles along the coast from Little Piskwamish Point, to the south, where they have been monitoring shorebirds. On a single day there, they counted 910 knots. With six people on Longridge, we can now cover the entire shore every day. Sturdee is a retired shop owner from Toronto for whom two weeks on James Bay will provide a welcome opportunity to learn the finer points of shorebird identification. His skills also extend to creative cabin repair when squirrels rip into the roof tarps. Keaveney and Vandermeulen, skilled naturalists, are both doing an Ontario Big Year, competing to see the most bird species in Ontario in a single year. They hope to see Arctic terns.

One day Charlton, Vandermeulen, and I walk five miles through a broad, shallow bay behind the ridge to meet birds coming in with the tide. Sandhill cranes that fly over the cabin at night, often wakening me with their loud, piercing cries, are often there during the day, standing in the flats inland from the ridge, their rust-stained plumage shining in the sun. At the far end of the bay, Charlton turns her binoculars and then her scope on a black dot perched on a distant rock—a black guillemot. Still in breeding plumage and with distinct white patches on its wings, this seabird is an unexpected find, and another species for the 22-year-old Vandermeulen. Recognizing a bird by its shape is second nature to Charlton. "If you see 16 people at night in silhouette," she tells me, "you'd recognize your daughter. It's the same thing." A few knots fly overhead.

On another day, while Iron and I walk the bay, dunlins gather. Pectoral sandpipers, birds with throat sacs that inflate during courtship, walk in the grass. We pause to smell fragrant wild sage. For Iron, birding is not only, or not even primarily, about lists. It's about aesthetics and the "intricate beauty of birds": the curve of flight feathers on a Kumlien's gull, the thrill of standing in a wide, open marsh, watching thousands of sandpipers passing overhead, or walking long ridges to approach birds that fly "impossible distances." She leads me right up to the pectoral sandpipers, Arctic breeders and champion long-distance fliers that will continue on to the pampas of southern South America, an annual round trip of 19,000 miles.

For Charlton and for her, it's also about the pressing need to count birds, to record their comings and goings. They find great joy in this work, volunteering months of their time. "We need to document the passage of birds through this wilderness," Iron says, "to protect them before it's too late." Iron brings James Bay to the outside world through her website, show-casing close-up photographs taken by using the lid of a spice jar to attach her camera to her spotting scope. She also shares her experiences with Ontario Field Ornithologists, a 1,000-member organization dedicated to the study of Ontario bird life, of which she was president for nine years. She dreams that the data she and others give so much of their time to collect will earn the hidden but spectacular bird sanctuary in James Bay the inter-national recognition it needs and deserves, perhaps as part of the Western Hemisphere Shorebird Reserve Network.

In the mornings and evenings, sitting outside eating breakfast or dinner, we watch white-winged crossbills up in spruce trees prying open pinecones to get at the seeds. So many birds are opening pinecones we hear the seed coats and scales hitting the ground. We listen for the tick, tick calls of rails. On most days while walking the ridge, I am surprised to see large flocks of European starlings, 450 birds, wheeling above the trees. Introduced into the United States in 1890 and 1891 from a flock of approximately 100 birds released in Central Park, they now number some 200 million birds, "arguably the most successful avian introduction on this continent," spreading beyond North America's crowded cities, roaming this landscape few people frequent. I am seeing a lot of birds, a lot of large wildlife, and few knots.

As Vandermeulen and I head out into the meadow early one morning, we see a gray wolf loping leisurely out to the ridge. We take the same route as the wolf, whose tracks we continue to see in the days to come. At the far end of the point, we see flightless mergansers in the midst of molting, rafts of common goldeneye, and black scoters. In the blink of an eye, two juvenile marbled godwits alight on the rocks, and then just as quickly depart. We could easily spend hour after hour here, but we need to time our return so we're not stranded by rushing tides.

Another early morning, around 6:00 a.m., Charlton and I walk the mouth of the creek, less than a mile from camp, looking at Bonaparte's gulls. They nest here, a bit inland, in spruces overlooking the bogs—the only gull in North America to nest regularly in trees. Charlton spots a tern standing on a rock, its short red legs barely visible and decidedly shorter than those of common terns nearby. We radio the camp. Vandermeulen races down. The arctic tern takes off before he arrives, but he has another chance early the next evening, when it circles the mouth of the creek. The team also sees the world's smallest gull, the little gull, whose primary nesting area in North America may be the muskeg and sedge wetlands off the western coast of James Bay. At Longridge, Iron has watched young little gulls begging food from their parents, a sign they'd been nesting nearby.

In the coming years, the vast, undisturbed reaches of the Hudson Bay Lowland may undergo dramatic change as Canada begins to mine its rich mineral deposits. Between 100 and 200 million years ago, a plume of hot melted rock rising from deep within Earth's mantle ignited volcanoes that carried diamonds up toward the surface. The De Beers diamond company extracts them today, about 60 miles west of James Bay. About 200 miles west of the bay, rock more than 2.5 billion years old, dating from a time when the planet's first continents were built, contains rich mineral lodes of chromite, gold, nickel, and copper known as the Ring of Fire. Its potential for generating wealth has been likened to the Alberta oil sands. Peatlands, like those in the Hudson Bay Lowland, cover only 5 percent of Earth's land, but store 25 percent of its carbon. Opening the Ring of Fire, building the accompanying roads and infrastructure, and tearing apart one of the world's largest remaining wetlands may leave an imprint long after the mines themselves have closed.

Twelve major rivers flow from the Hudson Bay Lowland into Hudson and James bays, diluting the bays' salinity to one-third that of the sea. Water would naturally pour in as snow melts in the spring, slacken in summer and fall, but now the rhythm of rivers is timed to electricity demand in urban Ontario and Québec and even New England. On a clear day we can see the transmission lines. Twenty-nine hydroelectric generating structures line the Moose River and its tributaries alone. Knots were staging on the marshy islands, sandy beaches, and tidal mudflats of Hudson Bay's western shores long before the dams were built. In the spring of 1974, 3,500 knots were seen at Churchill (the Churchill River diversion was completed in 1977). How rerouting the Albany and Churchill rivers and altering salinity and circulation touches the lives of millions of migrating and nesting birds, and diminishes or enhances the food they eat is unknown. Forty percent of the Churchill's flow has been diverted to dams along the Nelson River, and the headwaters of the Albany River have been redirected into the Nelson and into the St. Lawrence, out of the Hudson Bay watershed altogether.

Data downloaded from geolocators fastened to knots in Texas and Massachusetts show knots still refueling along the Nelson and Hayes rivers, in both spring and fall. Ann McKellar, wildlife biologist with Environment Canada, is documenting the passage of knots through here, using new ears on the ground made possible by technological advances and declining prices of radio telemetry. University and government scientists, collaborating in a massive undertaking led by Phil Taylor of Acadia University and under the aegis of Bird Studies Canada, are beginning a project that will eventually employ hundreds of radio antennas and receivers across Canada and down through the eastern seaboard of the United States: along the Great Lakes, from Toronto out to the Maritimes, including the Bay of Fundy and the Mingan Islands; in the Arctic, on Southampton and Coats islands; on coastal Hudson Bay near the Nelson River; at North Point, Longridge, and Piskwamish on James Bay; and along the Gulf of Maine and down through Cape Cod.

Fully automated, solar-powered receivers equipped with credit card–sized Linux beaglebone computers record signals from radio-tagged birds flying within six miles of the towers. The computers can telephone the data into servers where researchers download them onto their computers. In the first trial of the new array, in the spring of 2014, McKellar picked up radio signals from 10 knots tagged a month earlier in Delaware Bay. She also looked for shorebirds by helicopter. In an area near the Hayes River, near where Andrew Graham and Thomas Hutchins may have sighted their knot in 1770, she found knots feeding in salt marshes behind tidal flats still covered in ice. She saw more knots than any shorebird except semipalmated sandpipers—as many as 1,900 birds at one time. As the study continues, who knows what additional key pieces of the knot migration trail will be revealed?

On the long ride back to Toronto, the birders continue counting. They are accustomed to keeping lists. In addition to life lists and yearly Ontario lists, they track birds in their backyards, on television, and even bird abbreviations on license plates. ("AMAV" on a plate gets credited as "American Avocet.") This time they count how many birds they see every six miles. Overall, they log in 2,788 individual birds and 30 species, including hawks and harriers, kestrels and a kinglet, cedar waxwings and sandhill cranes. By the end of the year, Vandermeulen will finish his Ontario Big Year with a record-breaking 344 species, five of which—red knot, Nelson's sparrow, black guillemot, arctic tern, and buff-breasted sandpiper—he saw only in James Bay. He found 118 species at man-made reservoirs and 151 species, including 26 ducks and 22 shorebirds, at sewage lagoons. He found some of the rarest species—the gray-crowned rosy finch and the band-tailed pigeon—at bird feeders.

Vandermeulen found 307 of his 344 species on his own traveling by himself. He learned about the remaining 37 from the "Ontbirds [Ontario birds] listserv, text messages, e-mails, phone calls, and the database eBird (www. ebird.org)." Receiving an e-mail from Ontbirds about a sighting of a rare Bell's vireo in Point Pelee National Park—recorded only 12 times previously in Ontario—he raced over to find it. An e-mail from a friend sent him to his last species of the year, a Pacific loon.

The Cornell Laboratory of Ornithology and the National Audubon Society launched eBird in 2002. The site has grown 40 percent per year since its inception, with thousands of birders contributing millions of observations that so far include 96 percent of the world's 10,324 species. The Burrell brothers submit their sightings to eBird; one day, perhaps, they will provide a historical record that will help birds at risk. In the meantime, birders who report their sightings at sewage lagoons, prairie potholes, or inlets where tidal creeks wash into the sea may alert the rest of the world to unrecognized places where knots take refuge.

During the rest of our stay on James Bay, we see very few knots on their way south. Our peak counts, and those at other James Bay camps, are approximately half what they've been in previous years. It appears to have been a hard summer, not only in East Bay but in other Arctic knot nesting areas as well. When trouble can originate in so many places along the flyway, it's hard to know how much to attribute low numbers here to birds not finding enough horseshoe crabs in Delaware Bay, to the erratic weather in Southampton, to the dearth of lemmings, which may have sent hungry foxes to shorebird nests, or to a disturbance elsewhere. If each rung along the ladder is made stronger, the number of knots should become large enough to survive a bad year.

Many, but not all, knots leaving the Arctic build up their energy reserves in James Bay. Ornithologists canoeing its shores recognized the bay's importance to shorebirds 60 years ago, but scientists have now pinpointed another stopping area, an island archipelago in one of the world's largest estuaries, the Gulf of St. Lawrence. I'd been to Québec's Mingan Islands 10 years before scientists began counting knots there to look for Earth's largest animal, the blue whale, a long-distance traveler that occasionally swims up the gulf for a few days in the summer. We'd broken up the long days chasing whales, usually minkes, with brief sojourns on the islands. On Île Nue—the bare island—we ate lunch amid broken terra cotta tile and shards of burned barrel staves, remnants of a Basque tryworks from the 1500s, when whalers built ovens from sea clay and boiled whale blubber for its precious oil. Hiking along the shore in search of colorful puffins, I walked by shorebirds, unseeing. Seeking large, charismatic, long-distance travelers, I missed the tiny ones before me, whose journeys are no less momentous.

Biologist Yves Aubry from the Canadian Wildlife Service read historical reports and combed through 5 million records of Québec bird sightings dating back to the 1950s. He hired naturalists Christophe Buidin and Yann Rochepault, who knew the area and knew shorebirds, to scour the St. Lawrence coast looking for *bécassaux maubèches*, red knots. In 2006, they found 500 feeding at the limestone edges of islands in Canada's Mingan Archipelago National Park Reserve. Of the park's 40 small islands and 1,000 islets, knots prefer a few: Île Nue; Grand Île; Île Quarry, with its abundance of stone; and Île Niapiskau, the "waiting-for-the-ducks" island, named by Innu who once lived here.

During the summer field season, Aubry rents a house overlooking the river in the town of Mingan. Each day, the field crew and he catch Parks Canada boats to the islands, old boats that have seen better days. At the helm, the ever-cheerful Pierrot Vaillancourt nurses a tired engine that tends to overheat. In addition to his mechanical skills, Captain Vaillancourt navigates through thick fog, to drop-offs materializing only as he glides in to shore. The team splits up, each member to a different island. The fog, produced by warm, windless air resting on cold water, can last for weeks. No one is deterred. Aubry and I walk through a wood fragrant with balsam, listening to waves in the distance, following the voices of singing shorebirds. Eventually he finds them, a flock of darkened silhouettes at the edge of a shallow reef. We edge toward the birds, painstakingly slowly. We wear drab colors, keep our scopes pointed down and, to minimize our obvious presence, I follow directly behind Aubry. The water occasionally tops my waders. The rock is slippery. I focus on not dropping the scope that's been entrusted to me.

Some 450 million years ago, rivers dropped sand and silt into a tropical sea, and shells of animals long gone—sea lilies and sponges, coral and mollusks—slowly accumulated, building a plateau two miles thick. As continents drifted and ocean basins opened and closed, the plateau moved north. Seventy million years ago, it rose from the sea, to be cut and eroded by rivers, leaving the Mingan Islands. Glaciers advanced and retreated, sea level rose and fell, and today wind, wave, and winter frost wear away what remains, carving giant curved sculptures from the rock. These "flowerpots" tower above us, but they too will wash away. We climb over one that has already collapsed. In brief interludes when sunlight penetrates the fog, we can see puffins skidding into the water. Their nesting areas no longer disturbed by human intrusion, their numbers have steadily increased, from 163 pairs in 1985 to 960 pairs in 2005 to well over 1,000 pairs in 2012.

We find bécassaux maubèches where raised gravel barrens slope gently into water paved with ancient limestone. The birds are jumpy. The wind shifts ever so slightly, and they are gone. Slowly, we follow. Again, Aubry finds them in the white stillness and again, we wait. Startled by something we can't perceive, they scatter. The hours stretch before us. Aubry is patient. In his basement in Québec, he has lovingly cared for 5,000 orchids. A judge for the American Orchid Society, he discovered an orchid in Cuba that was named for him, *Lepanthus aubryi.* Late in the afternoon, when we've returned to the house, beautiful flute music emanates from his room. He'd trained for many years to become a concert flutist but, coming to dislike the pressure of performing perfectly for large audiences, chose music as an avocation, where he can relax and more privately enjoy its pleasures. "Sometimes I like to be invisible," he says. Perhaps this is an appeal of fog. Unlike other scientists I accompany, he prefers not to live on adrenalin, almonds, and energy bars. We break for Black Forest ham sandwiches, cara cara oranges, and gourmet dark chocolate.

After lunch, we resume searching in the fog, Aubry gently calling to the birds. The light momentarily grows brighter, and he finds them, bécassaux maubèches and *tournepierres à collier*, turnstones. The turnstones are

Red knot juvenile (*left*) and adult (drawing by Michael DiGiorgio).

flipping seaweed; the knots are foraging on amphipods, periwinkles, and tiny mussels. The adults, losing their breeding plumage, look scruffy. Another foggy, rainy day, Aubry locates 250 knots, 40 of them juveniles with still-shining, fresh plumage. The last time Aubry saw large numbers of adult knots here was 2009, when many flocks contained more than 1,000 birds, and one flock 4,000. In 2011, his team and he counted 600 juveniles. Since then, in 2012 and 2013, there've been very few. On the way back to the house, in a bog by the side of the road, we see a red-throated loon swimming with two babies.

Aubry is well liked in Mingan. In the evenings, the field crew and his friends gather for elegant meals and animated conversation in French. Amélie Robillard and Ilya Klvana, author of *Coureur des bois*, about his solo trans-Canada crossing by kayak from British Columbia to the Viking settlement in Newfoundland's L'anse aux Meadows, prepare many of the dishes. We eat common eider, its French name, *eider à duvet*, synonymous with a down-filled quilt; gray seal, called *tête-de-cheval* for its long, horselike face; lobster mushroom paté; and for dessert, freshly picked cloudberries. Captain Vaillancourt's mother has already gathered 68 quarts of the tiny berries. Jars of canned seal fill Robillard's and Klvana's cupboards. Their freezer is packed with capelin. The next evening: razor and surf clams, turbot, mushroom and snowshoe hare dip, smoked capelin, scallops with béchamel sauce, and a homemade blueberry and lingonberry tart. Robillard and Klvana foraged or shot most of the ingredients for both dinners.

Food is plentiful. Robillard's and Klvana's larders are full. Some of this abundance may not last. The sea, absorbing 25 percent of human carbon dioxide emissions, is growing more acidic, its acidity increasing 26 percent since the Industrial Revolution and projected, if current fossil fuel emissions continue unabated, to increase by 170 percent by the end of the century, a rate of carbon dioxide emission 10 times faster than any time in the last 55 million years and perhaps unprecedented in the last 300 million. Mollusks are highly sensitive to increasingly corrosive seawater which, undersaturated in minerals necessary to build strong shells, stymies their growth.

Plumes of carbon dioxide–laden water on the Pacific coast curtailing the growth of seed oysters at Oregon's Whiskey Creek shellfish hatchery and threatening the industry with collapse offer a glimpse of the future. Acidic water is believed to have killed 10 million seed scallops, shutting down a hatchery in British Columbia. Larval clams and bay scallops suffer from ocean osteoporosis as well: those grown in waters with preindustrial levels of carbon dioxide live longer, grow twice as quickly, and their shells are thicker and more robust.

Wherever knots stop along the flyway—in Chile, Argentina, Brazil, Texas, Florida, Georgia, South Carolina, Virginia, New Jersey, Massachusetts, James Bay, and the Mingan Islands—they eat tiny clams and mussels whose viability will be threatened. Study after study shows larval mussels, their integrity degraded in seawater with anticipated pH, growing smaller, thinner, and weaker. Mussels cling to the rocks by tough threads that weaken in increasingly corrosive water. In addition, as ocean carbon dioxide levels rise, far fewer of the clam knots favor in James Bay survive to settle on the bottom, the declines ranging from 36 percent to 89 percent.

Colder waters, such as those off Tierra del Fuego, in the Gulf of St. Lawrence, and Hudson Bay, are expected to acidify sooner. Most surface water in the Arctic will be corrosive to shellfish within decades. A more acidic sea threatens an essential source of food for millions of shorebirds, millions of eiders whose diet is primarily shellfish, and the livelihoods of shell fishermen, who support an industry worth millions of dollars. A shore without shorebirds and a sea without clams, mussels, oysters, and scallops is an impoverishment painful to contemplate. For now, in the quiet waters of the Mingan Islands, ocean acidification is a hidden and still unrealized threat. After their stay on the Mingan Islands, turnstones are plump and round, "their skin stretched so thin," Aubry says, "it's like parchment, and I can see the fat beneath. Right now, we don't find lean birds late in the season."

Hundreds of millions of tons of silt pour into the Atlantic each year from the Amazon River, building the muddy coast and tidal flats of French Guiana, Guyana, and Suriname. Currents push silt and mud west, sometimes as much as a mile each year, slicing away and building up the coast. Mounding sand cuts mangrove swamps from the sea, leaving trees to rot in shallow lagoons where shorebirds feed and roost. This odd landscape is rich in food for shorebirds. In 1982, 1.75 million shorebirds wintered in the Guianas.

Knots, coming south from James Bay or Mingan, may pause here for as long as a month before flying on to more distant shores in Tierra del

Fuego. In 2011, a hunter from French Guiana handed a park ranger the band from a red knot shot in rice fields edging the sea. Before it was killed, people had seen this knot passing through Delaware Bay and Florida. In French Guiana, hunting is legal but unregulated. In August 2012, on a beach where currents erode mudflats, breach rice impoundments, and return land to the sea, Aubry, Dey, Niles, and volunteer Steve Gates counted 1,700 knots. Dikes crossing the mile-wide rice impoundments were littered with shotgun shells.

Aubry, counting knots on Mingan, working in collaboration with Parks Canada to ensure their safety there, is like a worried parent, knowing that protecting his children in their Canadian homes can't guarantee their safety in the broader world. Where shorebird hunting continues, they are vulnerable. In French Guiana, he is working with the University of Alaska and the U.S. Fish and Wildlife Service to identify who is killing shorebirds and why; with a local conservation organization, Le Groupe d'Étude et Protection les Oiseaux en Guyane, to create protected shorebird reserves; and with officials in France to protect migratory birds in their overseas departments, not only in French Guiana but also in Guadeloupe, where in September 2011, two migrating whimbrels were shot—Machi, who en route from Hudson Bay had detoured around a tropical storm and rested in Guadeloupe, and Goshen, who'd perhaps stopped in the same heavily hunted swamp.

Shorebirds are hunted in coastal Guyana and Suriname as well. In 1983, a researcher studying terns in Guyana watched four young boys who'd strung a wire across a field and whipped it up and down killing 55 shorebirds as they flew into it. The boys invited him to lunch: "strongly-tasting" red knot, stilt sandpiper, and yellow legs, chopped and fried with their bones and served with noodles.

In September 1978, Guy Morrison and Dutch biologist Arie Spaans found the first hard evidence of a James Bay–South America knot connection. Counting birds in the lagoons and coastal mudflats of Suriname, they sighted three knots Morrison and Brian Harrington had seen only 23 days earlier, 3,500 miles away, in James Bay. Shorebird hunting is illegal in Suriname, and reserves protect the coast, but hunters reported taking tens of thousands of birds every year. Spaans, once he spoke of this on Suriname television, heard more: 30,000 shorebirds taken in one year in one district of Suriname alone, and hundreds of wintering semipalmated sandpipers shot at each lunar high tide along a single stretch of beach.

With logistical support and funding from the New Jersey Audubon Society, the indefatigable Morrison and Ross, taking to the air once again in the winters of 2008–11, documented a 79 percent decline in semipalmated sandpipers along the Guianas since their first survey. An international effort is under way to protect Suriname's shorebirds, including Spaans, the Nature Conservation Department of Suriname's Forest Service, and New Jersey Audubon. Every year, they teach 2,000 schoolchildren along the coast about shorebirds and the threats they face. They distribute posters identifying birds and their habitats; erect billboards on the main trunk road alerting people to the country's hunting prohibitions; rebuild the camps of game wardens and supply them with seaworthy boats, motors, and gasoline. They hope that by inviting the public to consider the great distances the birds travel and Suriname's critical importance to their well-being, people will take pride in this tiny country's rich diversity of birds and offer them safe harbor on the long journey home.

EPILOGUE

Heading Home

The days are getting shorter, summer is fading, and knots are moving south. Thousands once came through Massachusetts in the fall; their numbers are far fewer now. Flying in from James Bay or the Mingan Islands, some will refuel at Cape Cod's Monomoy National Wildlife Refuge. On a warm fall day, I join refuge biologist Kate Iaquinto as she steers a boat across Chatham Harbor. She anchors it far from the beach in water that will have drained away when we leave. Gray seals are hauled out on the sand, resting in the sun. Colonists once took them for their skins and oil and soon decimated a population that Farley Mowatt, reviewing maps, charts, and written accounts along the Atlantic coast, believes could have been between 750,000 and 1 million.

Bounty hunting for seals ended in Massachusetts in 1962. Since 1972, they've been protected under the Marine Mammal Protection Act; 15,000 now live in the coastal waters of southeastern Massachusetts, and as many as 7,000 can be seen in the refuge throughout the year. Today, 200 or 300 crowd the low-lying sandbars, and others bob in the waves a few feet from shore. As their numbers grow, great white sharks—considered vulnerable to extinction by the IUCN—are gathering off Monomoy to dine on the plump, energy-rich seals. We're wading through water that seems too shallow for great whites, but we're not that far from what is now known as

Shark Cove and its new great white shark café, where scientists have been tagging the animals, hoping to learn the habits of these elusive, enigmatic ocean wanderers, our oceanic counterparts at the top of the food chain: I move quickly.

Once on the beach, we are looking for juvenile knots. They could be anywhere along eight miles of sand; researchers carrying cell phones and GPSs have spread out to find them. We walk by bones from a gray seal bleached by the sun. Knots once flocked on Cape Cod's marshes and beaches in "exceedingly large numbers," wrote George H. Mackay. "This was previous to 1850 and when the Cape Cod railroad was completed only to Sandwich. Often, when riding on the top of the stage coach . . . beyond this point, immense numbers of these birds could be seen." How many? Sportsman S. Hall Barrett of Malden estimated that "in old times," he saw as many as 25,000 knots on Cape Cod in one year; between 1885 and 1893 he counted only 500 birds altogether. He made his observations from Billingsgate, an island near Wellfleet that has since washed into the sea. Horseshoe crabs were once abundant on Cape Cod as well. Henry David Thoreau, visiting the cape, reported pigs feeding on the many horseshoe crabs, which he called saucepan fish. Flocks of knots once scurried along the beaches eating their eggs. Seals and sharks are returning to the waters of Monomoy, restoring a world that had disappeared. Perhaps someday clouds of knots will return as well: today we will be lucky if we see 250.

Iaquinto and I begin walking through a soft, waterlogged swale that suddenly no longer seems firm. There's an awful squelching sound as I try to pull my feet from the sand. We're at the Connection, a newly emerging, saturated shoal rejoining two long barrier islands. Hurricanes and storms divide and redivide the barrier beaches of Monomoy, a name derived from the Algonquian for "a mighty rush of water." Here the sand is returning, and nearby, another island is emerging, Minimoy, where knots have already been seen. Across the way, outside the refuge, a few summer camps have fallen into the sea.

We find coyote tracks in the new, rising dunes, and about 50 knots. They stand facing into the wind, their heads tucked down, one leg tucked up. Three groups of knots migrate through Monomoy in the fall: adults who pause briefly before continuing on to Argentina and Tierra del Fuego;

adults who tarry for at least two months, molting into their winter feathers before heading south; and juveniles. Larry Niles, working with Iaquinto and others from the refuge, placed geolocators on 40 knots molting in Monomoy. So far, the team has retrieved eight. The birds are wintering in Florida, on the barrier islands of Virginia and Cape Hatteras, and in South Carolina, Georgia, and Cuba.

Many juvenile knots fly all the way to Tierra del Fuego, but not all. At a time when knots are threatened, scientists are eager to find the winter homes of juveniles. David Newstead and his colleagues identified one important wintering home—Texas's Laguna Madre. The day after Iaquinto and I walk the beach looking for juveniles, we join Niles and the rest of the team, who have located a flock. We try to catch the birds to deploy more data loggers, but it's windy and the knots are flighty. After an entire morning waiting and coaxing birds toward the net, only to have them scatter, Niles calls off the catch. The following day, with better weather, it succeeds, and the team deploys geolocators on 54 juvenile knots. As the birds resume their migration, scientists don't know whether light levels recorded by the devices will show birds traveling along a route already known or suspected, or perhaps reveal a new home, still hidden from us.

Felicia Sanders and the team in South Carolina recapture one of these knots. The battery on the geolocator is corroded, but the British Antarctic Survey is able to retrieve the data, which, when mapped by Ron Porter, show the young knot flying south to winter in Cuba, and then, to everyone's surprise, flying into the Cape Romain National Wildlife Refuge for the summer. Where there is one, there will be others: during the spring and summer this knot lived in the refuge, Mary-Catherine Martin, who surveys shorebirds there, recorded between 1,000 and 1,500 knots each month. If these are young birds as well, a new summer home for juvenile knots has been uncovered.

I've traveled the length of the flyway following knots. Now, I want to be at home when they might be migrating through. Dr. Charles Wendell Townsend's *Birds of Essex County*, published in 1905, describes Coffin's, Ipswich, and Plum Island beaches, long, thin barrier beaches I have walked for many years but where I have never seen knots. His account of cod and haddock throwing themselves ashore by the thousands in pursuit of herring reads like fiction—an abundance I may never see. Many birds had been depleted when Townsend wrote. He rarely saw great blue heron or egret. On a fall day, I might kayak by six or seven great blue heron, and as many as 60 egret. He saw knots—rarely in the marsh, more often on the beach—in flocks of up to 12, where, he said, they were easily shot.

I find my first knots in Essex County on Plum Island in the Parker River Wildlife Refuge, not along the beach at the mouth of the river, where I thought they might be, and where on future walks I will find one or two, but rather in the man-made impoundments where managers manipulate water levels, releasing water to attract migrating shorebirds, exposing mud where they can roost and feed, and then allowing it to flow in again, attracting migrating ducks. We walk out on the dyke where, as the sun is setting, a friend points to the knots—between 30 and 40 standing in the calm, shallow water.

I'd never seen knots in the bay behind our house, but after watching so many researchers—Carmen Espoz in Chile, Patricia González in Argentina, David Newstead and Felicia Sanders in Texas and South Carolina, Paul Smith and the crew in East Bay, and Jean Iron and Yves Aubry in James Bay and the Mingan Islands—patiently walk so many hours and so many miles to find them, I was beginning to think that maybe I hadn't known how or where to look. My neighbors April Prita Manganiello and Derek Brown have been watching and counting birds in the bay for 15 years. If there are birds to be seen, Manganiello and Brown will find them. They know every shoal at every tide. I often see them in their canoe, crossing the bay to survey the flats, or swimming in the cold river, following the tide like the birds.

On an afternoon in early September, on an incoming tide, my husband and I paddle out through the marsh and creek where, so many years ago, I first watched horseshoe crabs coming in to spawn. There still aren't many, but from time to time I find the fresh, translucent casts of young molting crabs left behind on the sandbars or strewn in the fallen marsh grass. As time passes and my own losses mount, I find comfort in their pale, perfectly formed shells, each curved and clean and with a barely perceptible seam where a young crab left the old shell behind to walk into a still-fertile, still-nourishing sea, evidence that a web, once torn, can be respun.

Manganiello and Brown have anchored their canoe in the sand and waded up a shallow rivulet. When we catch up to them, they are gazing into the fading marsh grass. As the water rises, sandpipers, plovers, terns, and dowitchers—birds I've been watching through four seasons on two continents—are flying in, filling the roost. Manganiello counts 80 handsome black-bellied plovers and 300 semipalmated plovers.

The tide isn't particularly high, nine feet, but wind from a hurricane to the south pushes in water. The bay fills with white caps and big waves, but the marsh absorbs the chop. Where we are, shielded from the rough water, it is quiet and calm. I don't know how long we've been here—maybe a few hours—when over a slight rise my husband sees three juvenile knots, their feathers clean and crisp, shining in the late afternoon light. Manganiello and Brown count eight more. Eleven red knots born in that awful Arctic summer who survived. Were they hatched on a windswept, freezing ridge on Southampton Island while I was wrapped in layers of down? Or did their parents nest farther north, maybe on the Melville Peninsula, where Captain Lyon found the first knot nest on Quilliam Creek, or farther west, on the harsh summit of Victoria Island's Mount Pelly where scientists observed courting knots more than 50 years ago?

Here they are, 1,800 miles later, looking none the worse for wear. Where are they going? They may take small hops down the coast, stopping or perhaps wintering in Virginia, or on Kiawah Island, South Carolina, or maybe on Pritchard's Island, where Sanders, Segars, and I gazed out through the fog one November day at 1,000 sandpipers that looked as if they might be knots. Perhaps these young birds are going to Georgia to the mouth of the Altamaha River, to the beaches of St. Petersburg, Florida, or to Cuba. Perhaps they are going the distance, heading another 3,000 miles to French Guiana to refuel in mudflats near the rice fields, before making that last 4,500-mile flight to Tierra del Fuego.

Will these birds have fair winds? Will they wait for the hurricane to subside? Almost all major Atlantic hurricanes, 96 percent, occur between August and October. On this same September day a few years ago, one knot with a data logger, 1VL, left Massachusetts on a six-day nonstop flight to northern Brazil. The flight included a 600-mile detour back north to avoid high winds. Will these birds make it? Will they find their way to Bahía Lomas? Odds are a few may not. For those that do, perhaps Boris Cvitanic, tending sheep on his estancia, will be there to greet them.

The story of red knots begins with loss—loss of large numbers of birds, loss of beach and mudflat, loss of horseshoe crab eggs, and a slide toward extinction. As I began to understand this story, a close friend became seriously ill. Each time we were together, she asked after the knots and their seaside homes: the crowded beaches in Delaware, the gleaming oyster banks overrun with horseshoe crabs in South Carolina, the touristed beaches in Las Grutas. She watched the story as it unfolded, drawn to the tiny birds flying such great distances. I called her from the bottom of the world, when clouds of knots were swirling in the skies over Bahía Lomas, and from the top, from Iqaluit, a city whose name I couldn't pronounce correctly until I got there, and where, while knots were flying, we were grounded in a blistering storm. Before I left for the Arctic, she wanted to talk about how the story might end and what it all meant: she knew she wouldn't live to read the book. Sitting in her bedroom overlooking the creek filled with the tide, we gazed at each other in silence as I faced the reality of her departure. "Keep flying," I said finally, "and you will find your way home."

To whom was I speaking, and about what? To my friend on a hard journey down a road she'd never been to a place she didn't know? To her family, who would soon shape their lives, their routes, holding a heartbreaking loss? To myself?

The story of the red knot is a story of loss that turns toward restoration and renewal. It is a story of the tenacity and resilience of birds under terrible pressure making long journeys year after year, even as their homes are diminished and their food grows scarce. As we lose our own bearings, their long flights offer a compass. Flying from one home to the next, they carry an imprint of each, the quality of their lives in one home enhanced or diminished by their lives in another. At the end of their journey, they have taken the measure of a shoreline running the length of the Earth. Whether, in each of their homes between Tierra del Fuego and the Arctic, they will find shelter on quiet and spacious beaches, food in abundance, and marshes where they can roost, hangs in abeyance. We choose the chemistry and pH of seawater, the reach of tides, the resiliency of the sea edge.

The story of red knots is also the story of the tenacity and endurance of the many people who, year after year on beach after beach, give knots safe harbor and ensure they are well provisioned. Standing for red knots, they are standing for all shorebirds whose homes are shrinking. Standing for red knots, they are standing for horseshoe crabs, and for the many inhabitants of sea and shore whose lives, like ours, depend on these ancient animals. Protecting the red knot, they renew the web of life at the edge of the sea, repair our torn world. Time after time, I have watched their undaunted courage giving voice to the birds and horseshoe crabs that cannot speak. They need us now.

We can cease taking horseshoe crabs for bait, as South Carolina has, and we can require the use of newer alternative bait. We can lift the veil of secrecy surrounding the biomedical industry to lower the number of horseshoe crabs dying or failing to spawn when they are caught and bled. We can insist on cars and trucks with better mileage, power plants that burn less coal, and buildings that are far more efficient. We can renew the call for an excise tax on outdoor recreational equipment, similar to that on guns and ammunition, to protect the homes of birds we don't shoot as hunters protect the homes of those we do. Finally, we can offer horseshoe crabs and shorebirds—and ultimately ourselves—breathing room, by giving beaches freedom to migrate inland unimpeded as the water rises. Then, a bird whose very existence is now imperiled may once again, and perhaps in my lifetime, be safe and free, and ancient horseshoe crabs, enduring since the dawn of animal life, may once again thrive.

In the rush of wings as thousands of birds lift into a night sky, in the gentle voices of knots singing in the Arctic stillness, and in the emergence of horseshoe crabs on a quiet moonlit beach, beauty and compassion abide, and call. I remember the words of Maria Belén Pérez. "To hold a red knot and feel its beating heart," she'd told me, "is to feel the heartbeat of the Earth," an Earth we all share. Their home is ours. We stand together, all of us, on the edge, facing a time fraught with challenge, filled with promise.

Notes

ONE The "Uttermost Part of the Earth"

On the origin of the name "red knot," see Phillips, *The New World of Words*.

Records of knots wintering in South America include DeVillers and Terschuren, "Some Distributional Records"; Johnson, *Birds of Chile* ("among," 344); Meyer de Schauensee, *Species of Birds of South America;* Wetmore, *Our Migrant Shorebirds*.

On findings from the Harrington and Morrison road trip and the Morrison and Ross aerial surveys, see Harrington and Morrison, "Notes on the Wintering Areas of Red Knot"; Morrison and Ross, *Atlas of Nearctic Shorebirds*.

On Bahía Inutil, see King, *Voyages of the* Adventure *and* Beagle ("flattered ourselves," 124; "neither anchorage nor shelter" and "lost no time in retreating," 125).

On geographic dispersal of knots, see Buehler and Baker, "Population Divergence Times"; and Buehler, Baker, and Piersma, "Reconstructing Palaeoflyways."

For animal extinction and settlement in South America, see Barnosky and Lindsey, "Megafaunal Extinction"; Cione, Tonni, and Soibelzon, "Did Humans Cause?"; Latorre et al., "Late Quaternary Environments"; and Salemme and Miotti, "Archeological Hunter-Gatherer" ("It was a slow," 473).

On Magellan and other navigators, see Bergreen, *Over the Edge;* Morrison, *The European Discovery of America* ("23 charts" and "wine, olive oil, vinegar and beans," 343–44); and Slocum, *Sailing* ("struck like a shot").

On birds flying through or around hurricanes, Niles et al., "First Results"; Fletcher Smith, Center for Conservation Biology, personal communication, July 26, 2014; and Watts et al., "Whimbrel Tracking."

TWO When Is the Beginning of the End?

For passenger pigeons, see Audubon and Macgillivray, *Ornithological Biography*, vol. 1; Forbush, *Game Birds* ("Sunne never sees," 435); and Greenberg, *A Feathered River*.

On the great auk, see Newton, "Wolley's Researches" ("much less time than it takes to tell," 391); Townsend, *Birds of Essex County;* and Tuck, *People of Port au Choix*.

For more information on the status and search for ivory-billed woodpeckers, see www.iucnredlist.org; and http://www.birds.cornell.edu/ivory.

On spoon-billed sandpipers, see BirdLife International, "*Eurynorhynchus Pygmeus*"; Vyn, "Spoon-Billed Sandpiper"; Wildfowl and Wetlands Trust, "Saving the Spoon-Billed Sandpiper"; and Zöckler et al., "Hunting in Myanmar."

For the Texas whooping crane decision, see Jack, "Opinion and Verdict of the Court"; and *The Aransas Project v Shaw*.

For knot population trends, see Andres et al., "Population Estimates"; Carmona et al., "Use of Saltworks"; Summers, Underhill, and Waltner, "Dispersion of Red Knots"; Wetlands International, "*Calidris Canutus*"; and Yang et al., "Impacts of Tidal Land Reclamation."

For proposed listing in the United States, see U.S. Fish and Wildlife Service, "Proposed Threatened Status."

For the Eskimo curlew, see Cornell Lab of Ornithology, "Eskimo Curlew"; Forbush, *Game Birds* ("perhaps . . . bells" and "to denote," 418–19); Gill, Canevari, and Iverson, "Eskimo Curlew" ("numbered at least in the hundreds of thousands"); and U.S. Fish and Wildlife Service, *Eskimo Curlew*.

For the history of settlement and development along the Strait of Magellan, see Baldi et al., "Guanaco Management"; Martinic, *Brief History;* Morris, *Strait;* and Morrison, *European Discovery of America*.

For the history of Cerro Sombrero, see Bastidas, *Cerro Sombrero* ("la realización," 31).

For oil spills and birds in the strait, see "Berge Nice"; Flores, *Antecedentes sobre la avifauna;* Hann, *VLCC* Metula (Almost," 212); Hann, "Fate of Oil"; and Owens, "Time Series."

For impacts of oil spills on shorebirds, see Henkel, Sigel, and Taylor, "Large-Scale Impacts."

For more information about protecting shorebirds in Bahía Lomas, see https://www.facebook.com/pages/Centro-Bahia-Lomas/270509379698671.

For knots in Río Grande, see Escudero et al., "Foraging Conditions."

THREE The Urban Bird and the Resort

For Magellanic plover, see Ferrari, Imberti, and Albrieu, "Magellanic Plovers"; and Jehl, "*Pluvianellus Socialis*."

For the Brownsville dump, see Obmascik, *The Big Year*.

For Fresh Kills Landfill and Hurricane Sandy, see Kimmelman, "Former Landfill."

For more about shorebirds in Río Gallegos, see Río Gallegos Western Hemisphere Shorebird Reserve Network, http://www.whsrn.org/site-profile/rio-gallegos-estuary.

Numbers of knots in Río Gallegos: unpublished data from Silvia Ferrari and Carlos Albrieu.

For Ferrari and Albrieu's work in Río Gallegos, see Albrieu and Ferrari, "Participación de los municipios"; Albrieu, Ferrari, and Montero, "Investigación, educación e transferencia"; and Ferrari, Ercolano, and Albrieu, "Pérdida de hábitat."

For science and advocacy, see Runkle, *Advocacy in Science.*

To read more about challenges to scientists in the United States, see Mann, *The Hockey Stick;* Michaels, *Doubt Is Their Product;* and Oreskes and Conway, *Merchants of Doubt.*

For the hooded grebe, see BirdLife International, "*Podiceps Gallardoi*"; and Ambiente Sur at http://www.ambientesur.org.ar/.

For "hot legs," see Piersma and van Gils, *Flexible Phenotype.*

For feeding knots, see González, Piersma, and Verkuil, "Food, Feeding, and Refueling"; González, "Las aves migratorias"; and Piersma and van Gils, *Flexible Phenotype.*

For the soda factory, Di Giácomo, "Fabrica de soda solvay," unpublished report; Giaccardi and Reyes, *Plan de manejo;* Jenkins et al., *Brine Discharges;* and Diego Luzzatto, Consejo Nacional de Investigaciones Científicas y Técnicas (CONICET), unpublished data.

For more information on Inalafquen, see https://www.facebook.com/pages/Fundacion-Inalafquen/150422954977075.

For Darwin in Patagonia, see Darwin, *Voyage of H.M.S.* Beagle ("giant's bones" and "eighteen pence," 155; "wonderful relationship," 173; "cooked and eaten" and "Fortunately the head," 92–93; and "expand their wings," 90); and *Origin of Species* ("mystery of mysteries," 1; and "serve to show," 200).

For epigenetics, see Emerson, "Epigenetics"; Manikkam et al., "Epigenetic Transgenerational Inheritance"; Moczek et al., "Developmental Plasticity"; Nätt et al., "Inheritance of Acquired Behaviour"; Ng et al., "Chronic High-Fat Diet"; Nilsson et al., "Epigenetic Transgenerational Inheritance"; Richards, "Inherited Epigenetic Variation"; Saey, "From Great Grandma to You"; Szyf, "Lamarck Revisited"; and West-Eberhard, "Developmental Plasticity."

For epigenetics and flexibility in knots, see Piersma, "Flyway Evolution Is Too Fast"; Piersma and van Gils, *Flexible Phenotype;* and van Gils et al., "Gizzard Sizes."

For Darwin and extinction, see Darwin, *Voyage of H.M.S.* Beagle, 174.

Information about L6U from González, Niles, Watts, Kalasz, and Dey, and www.band edbirds.org, where stops of other flagged shorebirds can be found as well.

For more information on the work of Rare, see http://www.rare.org/.

For godwits, see Gill et al., "Hemispheric-Scale Wind."

Information about H3H arriving in Florida from Patricia González and Doris and Patrick Leary.

See Saint-Exupéry, *Night Flight* ("snugly ensconced," 8; "vast anchorage," 3; "deeply meditative," 10; "worms in a fruit," 5; and "fatal lure," 146).

FOUR Bay of Plenty

For discovery of Delaware Bay as an avian Serengeti, see Dunne et al., "Aerial Surveys"; Dunne, *Bayshore Summer* ("awash in birds," 18; and "than were estimated," 19);

Dunne, *Tales* ("no stranger to numbers of birds" and "like storm clouds," 12).
Myers, "Sex and Gluttony" ("sex and gluttony," 68; and "no other spot," 74).

For YoY and 1VL, see Niles et al., "First Results."

I used these field guides often: Kaufman, *Lives of North American Birds;* O'Brien,
Crossley, and Karlson, *Shorebird Guide;* Sibley, *Field Guide;* and Stokes and Stokes,
Beginner's Guide.

For nineteenth-century Delaware Bay, see Audubon and Macgillivray, *Ornithological
Biography,* vol. 3 ("laden with fish and fowls," 606); Audubon and Macgillivray,
Ornithological Biography, vol. 4 ("immense number," 123); Beesley, "Sketch of the
Early History" ("isolated as it was," 129); Cantwell, *Alexander Wilson;* "From Cape
May"; Wilson, *Wilson's American Ornithology* ("great multitudes" and "the
remains," 656; "bushels," "lying in hollows," 481; "almost wholly on the eggs," 480);
and Wilson, *Life and Letters.*

For preferred food in Delaware Bay, see Botton and Harrington, "Synchronies."

For twentieth-century shorebird sightings in New Jersey, see Potter, "The Season"
("thousands of shore-birds," 242); Shuster, "Natural History and Ecology of the
Horseshoe Crab"; Stone, *Bird Studies at Old Cape May* ("quotes an old," 400); and
Urner and Storer, "The Distribution and Abundance of Shorebirds."

On the history of horseshoe crabs and knots in Cape May County, Carole Mattessich
Raritz and J. P. Hand helped locate sources; "From Cape May" ("Old Salt," "utility
of king crabs," and "feed seabirds," 2).

For horseshoe crabs in Delaware Bay versus the Jersey ocean shore, see Fowler, "The
King Crab Fisheries in Delaware Bay"; and Rathbun, "Crustaceans, Worms,
Radiates, and Sponges."

For nineteenth-century abundance of horseshoe crabs, see New Jersey Geological
Survey, *Geology* ("so thick" and "shovelled up and collected," 106); and Wilson,
Wilson's American Ornithology ("their dead bodies," 481).

For the historical abundance of sturgeon in Delaware Bay, see Cobb, "The Sturgeon
Fishery of Delaware River and Bay"; and Saffron, *Caviar.*

For historical abundance of shark, see New Jersey Geological Survey, *Geology.*

For shad, see McDonald, "Fisheries of the Delaware River" ("finny race" and "planking,"
656); and McPhee, *The Founding Fish.*

For the oyster industry in Delaware Bay, see Hall, "Notes on the Oyster Industry of New
Jersey"; and Stainsby, *The Oyster Industry of New Jersey.*

See Wilson, *Wilson's American Ornithology* ("driven down, every spring," 481; and "egg-
nogg," "perfectly fresh," and "smelt abominably," 337).

For the horseshoe crab fertilizer industry, see New Jersey Geological Survey, *Geology;*
Rathbun, "Crustaceans, Worms, Radiates, and Sponges" ("a few years more," 830);
Smith, "Notes on the King-Crab Fishery of Delaware Bay" ("diminution in
the abundance," 366); and "The Great King Crab Invasion" ("to the probable
value" . . . "passed through a mill").

For shorebirds eating horseshoe crabs outside Delaware Bay, see Hapgood and Roosevelt,
Shorebirds ("have a *penchant*" and "poking out," 6); Forbush, *Game Birds* ("are

fond of the spawn," 267); Michael Haramis, unpublished records from the food habits archive, USGS Patuxent Wildlife Research Center, Laurel, Md.; and Sperry, *Food Habits of a Group of Shorebirds.*

For the absence of shorebird hunting on Delaware Bay, see Dunne, "Knot Then, Knot Now, Knot Later"; and Sutton, "An Ecological Tragedy on Delaware Bay" ("simply were not there," 32).

For naturalists hunting shorebirds, see Darwin, *Autobiography* ("in the latter part of my school life," 44); Pettingill, "In Memoriam" ("with every feather," 151); Sutton, "Birds of Southampton Island"; and "Parasitic Jaeger, Polar Bird of Prey, Seen Near Cape May" ("but as they had no arms," 13).

For names of knots, see Forbush, *Game Birds;* Hapgood and Roosevelt, *Shorebirds;* and Mackay, "Observations on the Knot."

For hunting shorebirds, see Fleckenstein, *Shorebird Decoys* ("countless numbers," "artistically," "nothing more," "flock after flock," 11–12); Forbush, *Game Birds* ("everybody shot," 264); Hapgood and Roosevelt, *Shorebirds* ("There are few more exciting experiences," 31); and Mackay, "Observations on the Knot."

For eating shorebirds, see Ball, *A History of the Study of Mathematics* ("plover; knottys" and "fesant in brase," 150); Mackay, "Observations on the Knot" ("only fair eating," 27); Thomas, *Delmonico's;* "Table Supplies and Economics"; and Fleckenstein, *Shorebird Decoys* ("hauled from the meadows," 13).

For loss of shorebirds and horseshoe crabs and partial recovery, see Bent, *Life Histories of North American Shore Birds* ("Excessive shooting," 132); Mackay, "Observations on the Knot" ("in a great measure have been killed off" and "are in great danger," 30); Shuster, "King Crab Fertilizer"; and Urner and Storer, "The Distribution and Abundance of Shorebirds" ("The increase in numbers," 193).

For former abundance of green sea turtles, see King, "Historical Review of the Decline of the Green Turtle"; and McClenachan, Jackson, and Newman, "Conservation Implications of Historic Sea Turtle Nesting Beach Loss."

For a 10 percent world, see MacKinnon, *Once and Future World.*

FIVE Tenacity

For rates that animals become fossils, see Prothero, *Evolution.*

For evolution of knot into its own species, see Baker, Pereira, and Paton, "Phylogenetic Relationships"; Gibson and Baker, "Multiple Gene Sequences"; and Jetz et al., "Global Diversity of Birds."

For the evolution of knots into today's lineages, see Baker, Piersma, and Rosenmeier, "Unraveling the Intraspecific Phylogeography"; and Buehler, Baker, and Piersma, "Reconstructing Palaeoflyways."

For the oldest horseshoe crabs, see Rudkin, "The Life and Times of the Earliest Horseshoe Crabs"; Rudkin, Young, and Nowlan, "The Oldest Horseshoe Crab"; Van Roy et al., "Ordovician Faunas of Burgess Shale Type"; and Young et al., "Exceptionally Preserved Late Ordovician Biotas."

For *Archaeopteryx* and horseshoe crabs in Solnhofen quarry, see Lomax and Racay, "A
    Long Mortichnial Trackway"; Wellnhofer, *Archaeopteryx* ("perfectly agrees with a
    bird's feather," 46); and "Palaeontology."
For knot survival rates, see Schwarzer et al., "Annual Survival of Red Knots."
For more about the Western Hemisphere Shorebird Reserve Network and Manomet, see
    Myers et al., "Conservation Strategy"; and http://www.whsrn.org/.
For a discussion of knot populations in Delaware Bay and elsewhere, and their decline,
    see Myers, "Sex and Gluttony" ("extraordinary concentrations," 73); and U.S. Fish
    and Wildlife Service, "Rufa Red Knot Ecology and Abundance."
For human disturbance, see Burger and Niles, "Closure versus Voluntary Avoidance."
For stranding crabs, see Botton and Loveland, "Reproductive Risk."
For more about "just flip 'em" and the work of the nonprofit Ecological Research and
    Development Group, see http://horseshoecrab.org/.
For knots in Virginia, see Barnes, Truitt, and Warner, *Seashore Chronicles* ("ten thou-
    sand," 113–14); Cohen et al., "Day and Night Foraging"; Duerr, Watts, and Smith,
    Population Dynamics of Red Knots; Jones, Lima, and Wethey, "Rising Environmental
    Temperatures"; Smith et al., *An Investigation of Stopover Ecology;* Barry Truitt,
    personal communication, April 14, 2013 ("two flags from Delaware Bay"); and U.S.
    Fish and Wildlife Service, "Rufa Red Knot Ecology and Abundance."
For eating habits of knots in Delaware Bay, see Atkinson et al., "Rates of Mass Gain and
    Energy Deposition"; Cohen et al., "Day and Night Foraging"; Haramis et al.,
    "Stable Isotope and Pen Feeding"; Mizrahi and Peters, "Relationships between
    Sandpipers and Horseshoe Crab"; Piersma and Gils, *Flexible Phenotype* ("shore-
    birds as a group have unrivalled capacities to process food and refuel fast," 74);
    Tsipoura and Burger, "Shorebird Diet"; and U.S. Fish and Wildlife Service, "Rufa
    Red Knot Ecology and Abundance."
For horseshoe crab egg trends in Delaware Bay and knot weight gains, see Baker et al.,
    "Rapid Population Decline in Red Knots"; Botton, "The Ecological Importance
    of Horseshoe Crabs"; Botton and Harrington, "Synchronies"; Botton, Loveland,
    and Jacobsen, "Site Selection by Migratory Shorebirds"; Dey, Kalasz, and
    Hernandez, "Delaware Bay Egg Survey, 2005–2010"; Mizrahi, Peters, and Hodgetts,
    "Energetic Condition of Semipalmated and Least Sandpipers"; Mizrahi and
    Peters, "Relationships between Sandpipers and Horseshoe Crab"; Smith, Millard,
    and Carmichael, "Comparative Status and Assessment of *Limulus*"; and U.S. Fish
    and Wildlife Service Shorebird Technical Committee, *Delaware Bay Shorebird–
    Horseshoe Crab Assessment Report.*
For declines in semipalmated sandpipers and ruddy turnstones, see Clark, Niles, and
    Burger, "Abundance and Distribution of Migrant Shorebirds"; Mizrahi, Peters, and
    Hodgetts, "Energetic Condition"; David Mizrahi, New Jersey Audubon, personal
    communication, April 4, 2014; Paul Smith, Environment Canada, personal
    communication, May 16, 2014.
For the decline in horseshoe crabs, see ASMFC Horseshoe Crab Stock Assessment
    Subcommittee, *2013 Horseshoe Crab Stock Assessment;* Davis, Berkson, and Kelly,
    "A Production Modeling Approach" ("trash fish," 215); Mizrahi, Peters, and

Hodgetts, "Energetic Condition"; and Mizrahi and Peters, "Relationships between Sandpipers and Horseshoe Crab."

For "whole stretches of beach," see Myers, "Sex and Gluttony," 74.

For early work to stem decline in horseshoe crabs, see Eagle, "Regulation of the Horseshoe Crab Fishery"; Loveland, "The Life History of Horseshoe Crabs"; Smith, Millard, and Carmichael, "Comparative Status and Assessment."

six Blue Bloods

For more on bacteria in the human body, see Qin et al., "A Human Gut"; and Specter, "Germs Are Us."

For the history of IV therapy, see Howard-Jones, "Cholera Therapy" ("benevolent homicide," 373; "carefully strained," 391); and "The Cholera" ("full of sound and fury, signifying nothing," 266).

For the work of Florence Seibert, see Rietschel and Westphal, "Endotoxin"; Rossiter, *Women Scientists in America;* Seibert, "Fever-Producing Substance"; and Seibert, *Pebbles on the Hill of a Scientist* ("seemed of moderate interest at the time" from Esmond Long in the foreword, vii).

For vision in horseshoe crabs, see Barlow and Powers, "Seeing at Night and Finding Mates" ("studying vision in a blind animal," 83; and "after many cold and lonely nights," 95).

For the history of the development of LAL, see Banerji and Spencer, "Febrile Response to Cerebrospinal Fluid Flow"; Cooper and Harbert, "Endotoxin as a Cause of Aseptic Meningitis"; Levin, "History of the Development of the Limulus Amebocyte Lysate Test"; Levin, Hochstein, and Novitsky, "Clotting Cells and Limulus Amebocyte Lysate"; Rietschel and Westphal, "Endotoxin" ("little blue devil," 1); and Thomas, *The Lives of a Cell* ("the very worst . . . shambles," 78–79).

For stone crab, see Goode, *The Fisheries and Fishery Industries of the United States,* section 1 ("by the hand," 773).

For more on the production of LAL, see Levin, Hochstein, and Novitsky, "Clotting Cells and Limulus Amebocyte Lysate"; Levin, "The History of the Development of the Limulus Amebocyte Lysate Test"; Novitsky, "Biomedical Applications of Limulus Amebocyte Lysate"; and Swann, "A Unique Medical Product (LAL) from the Horseshoe Crab."

For the gentamicin recall, see Fanning, Wassel, and Piazza-Hepp, "Pyrogenic Reactions to Gentamicin Therapy"; and Friedman, "Aseptic Processing Contamination Case Studies."

seven Counting

Recent knot population, egg density, and state of shorebirds from Amanda Dey, New Jersey Fish and Wildlife, and Larry Niles, LJ Niles Associates, personal communication, June 22, 2014; Dey et al., "Delaware Bay Horseshoe Crab Egg Survey, 2005–2012"; David Mizrahi, New Jersey Audubon, personal communication, June 16, 2014; and U.S. Fish and Wildlife Service, "Rufa Red Knot Ecology and Abundance."

For knot fitness after northeast storm, see Dey et al., "Delaware Bay Horseshoe Crab Egg Survey, 2005–2012."

For reverberations along the flyway, see Escudero et al., "Foraging Conditions 'at the End of the World' "; González, Baker, and Echave, "Annual Survival of Red Knots Using the San Antonio Oeste Stopover."

For horseshoe crab population trends, see ASMFC Horseshoe Crab Stock Assessment Subcommittee, *2013 Horseshoe Crab Stock Assessment;* Delaware Bay Ecoystem Technical Committee Report, "ARM Recommendation"; Horseshoe Crab Plan Review Team, *2013 Review of the Fishery Management Plan for Horseshoe Crab;* Smith et al., "Evaluating a Multispecies Adaptive Management Framework"; and U.S. Fish and Wildlife Service, "Proposed Threatened Status" ("stagnated," 60063).

For eel fishery, see ASMFC, *American Eel Benchmark Stock Assessment;* ASMFC, *Draft Addendum III;* Lane, "Eels and Their Utilization"; MacKenzie, "History of the Fisheries of Raritan Bay" ("chopped in half or quarters," 16); and Smith, Millard, and Carmichael, "Comparative Status and Assessment of *Limulus*."

For whelk fishery, see ASMFC Horseshoe Crab Stock Assessment Subcommittee, *2013 Horseshoe Crab Stock Assessment;* Fisher and Fisher, *The Use of Bait Bags;* and Horseshoe Crab Plan Review Team, *2013 Review of the Fishery Management Plan for Horseshoe Crab.*

For shad, sturgeon, and river herring, see ASMFC, *ASMFC River Herring Benchmark Assessment;* ASMFC, *American Shad Stock Assessment Report;* and NMFS, *Atlantic Sturgeon New York Bight Distinct Population.*

For alternative bait, see Fisher and Fisher, *The Use of Bait Bags;* Shuster, Botton, and Loveland, "Horseshoe Crab Conservation"; and Wakefield, *Saving the Horseshoe Crab.*

For threat of toxins and parasites from Asian horseshoe crabs, see Aieta and Oliveira, "Distribution, Prevalence, and Intensity of the Swim Bladder Parasite *Anguillicola Crassus*"; ASMFC, *ASMFC Approves Resolution to Ban the Import and Use of Asian Horseshoe Crabs;* Botton and Ito, "The Effects of Water Quality on Horseshoe Crab Embryos and Larvae"; Kanchanapongkul, "Tetrodotoxin Poisoning Following Ingestion of the Toxic Eggs of the Horseshoe Crab"; Kanchanapongkul and Krittayapoositpot, "An Epidemic of Tetrodotoxin Poisoning"; Leibovitz and Lewbart, "Diseases and Symbionts"; Machut and Limburg, "*Anguillicola Crassus* Infection"; Moser et al., "Infection of American Eels"; Muston, "Cafe de Mort"; Ngy et al., "Toxicity Assessment for the Horseshoe Crab"; Shin and Botton, letter to the U.S. National Invasive Species Council; Székely, Palstra, and Molnar, "Impact of the Swim-Bladder Parasite" ("serious threat for the overall reproductive success," 219); and U.S. Food and Drug Administration, *Bad Bug Book.*

For unaccounted horseshoe crab losses in the horseshoe crab fishery and losses in the biomedical industry, see ASMFC Horseshoe Crab Stock Assessment Subcommittee, *2013 Horseshoe Crab Stock Assessment Update* ("oversight" and "may account," 12); ASMFC Delaware Bay Ecosystem Technical Committee, *Meeting Summary* ("an accurate portrayal," 3); ASMFC Horseshoe Crab Technical Committee, *Meeting Summary* ("will eclipse" and "essentially equal," 1); Delancey and Floyd, *Tagging of Horseshoe Crabs;* Hurton, Berkson, and Smith, "The Effect of Hemolymph Extraction"; Kurz and James-Pirri, "The Impact of Biomedical Bleeding"; Leschen and Correia, "Mortality in Female Horseshoe Crabs"; and New Jersey Audubon et al., "Public Comments."

For proportion of male and female horseshoe crabs spawning on beaches, see James-Pirri, *Assessment of Spawning Horseshoe Crabs* ("extreme," 26); James-Pirri et al., "Spawning Densities, Egg Densities, Size Structure"; Rathbun, "Crustaceans, Worms, Radiates, and Sponges" ("in pairs," 829; and "it is not an uncommon thing," 829–30); and Smith, "Notes on the King-Crab Fishery" ("sometimes," "two or more males," and "seek the sandy shores," 363, 364).

For injuries caused by taking horseshoe crabs for bleeding, see Anderson, Watson, and Chabot, "Sublethal Behavioral and Physiological Effects"; Hurton, Berkson, and Smith, "The Effect of Hemolymph Extraction"; Kurz and James-Pirri, "The Impact of Biomedical Bleeding"; Leibovitz and Lewbart, "Diseases and Symbionts" ("traumatic injuries" and "stab-like wounds," 248); Leschen and Correia, "Response to Associates of Cape Cod"; Leschen and Correia, "Mortality in Female Horseshoe Crabs"; and Levin, Hochstein, and Novitsky, "Clotting Cells and Limulus Amebocyte Lysate."

For the horseshoe crab reserve, see ASMFC, *Addendum I to the Fishery Management Plan* ("taking of horseshoe crabs for any purpose," 5); NOAA, "Atlantic Coastal Fisheries Cooperative Management Act"; and Smith, Millard, and Carmichael, "Comparative Status and Assessment of *Limulus*" ("older juvenile and newly mature females," 367).

For rising demand for horseshoe crabs and declining Asian supply, see Botton et al., "Emerging Issues in Horseshoe Crab Conservation"; Chen and Hsieh, "The Challenges and Opportunities for Horseshoe Crab Conservation in Taiwan"; Dubczak, "Proven Biomedical Horseshoe Crab Conservation Initiatives"; Gauvry and Janke, "Current Horseshoe Crab Harvesting Practices" ("critical levels," PT-4); Hu et al., "Distribution, Abundance and Population Structure of Horseshoe Crabs"; and Seino, "A Reconsideration of Horseshoe Crab Conservation Methodology in Japan."

For the development of synthetic LAL, see Ding and Ho, "Endotoxin Detection"; Ding and Ho, "Strategy to Conserve Horseshoe Crabs"; Ding, Zhu, and Ho, "High-Performance Affinity Capture-Removal of Bacterial Pyrogen"; Levin, Hochstein, and Novitsky, "Clotting Cells and Limulus Amebocyte Lysate"; Loverock et al., "A Recombinant Factor C Procedure"; Sutton and Tirumalai, "Activities of the USP Microbiology and Sterility Assurance Expert Committee" ("important reason for revision," 10); and U.S. Food and Drug Administration, *Guidance for Industry.*

EIGHT Lowcountry

For sharks and loggerheads in Cape Romain, see Botton and Shuster, "Horseshoe Crabs in a Food Web"; Quattro, Driggers, and Grady, "*Sphyrna Gilberti* Sp. Nov., a New Hammerhead Shark"; and Ulrich et al., "Habitat Utilization."

Tiger sharks along the South Carolina coast and eating habits of sharks from Bell and Nichols, "Notes on the Food of Carolina Sharks"; Driggers et al., "Pupping Areas"; and William Driggers, National Oceanic and Atmosphere Administration, personal communication, July 23, 2014.

For long-billed curlew, see Andres et al., "Population Estimates"; and Audubon and Macgillivray, *Ornithological Biography*, vol. 3 ("The flocks enlarge," 242).

For history of knots in South Carolina, see Sprunt, "In Memoriam"; Sprunt and Chamberlain, *South Carolina Bird Life* ("an untrammeled wildness," 239); and Wayne, *Birds of South Carolina*.

For knots migrating through South Carolina more recently, see Given, "Leucistic Red Knot *Calidris Canutus*"; Michael Haramis, unpublished records from the food habits archive, USGS Patuxent Wildlife Research Center, Laurel, Md.; Leyrer et al., "Small-Scale Demographic Structure"; Marsh and Wilkinson, "Significance of the Central Coast of South Carolina"; Niles, Sanders, and Porter, unpublished data; Thibault, *Assessing Status and Use;* Thibault and Levisen, *Red Knot Prey Availability;* and U.S. Fish and Wildlife Service, "Rufa Red Knot Ecology and Abundance" ("acted as if they had perfect," 1227).

For history of knots and shorebirds eating horseshoe crab eggs, see Cape Romain National Wildlife Refuge, *Comprehensive Conservation Plan;* Riepe, "An Ancient Wonder of New York"; Rudloe, *The Wilderness Coast;* and Sperry, *Food Habits* ("fed almost exclusively on spawn," 14).

For oystercatchers on Marsh Island, see Sanders, Spinks, and Magarian, "American Oystercatcher."

For 2014 South Carolina horseshoe crab permit, see South Carolina Department of Natural Resources, "Horseshoe Crab Hand Harvest Permit HH14."

For horseshoe crabs and shorebirds at the Monomoy National Wildlife Refuge, see Anderson, Watson, and Chabot, "Sublethal Behavioral and Physiological Effects"; Eastern Massachusetts National Wildlife Refuge Complex, *Compatibility Determination* ("inviolate sanctuary ... for migratory birds"); James-Pirri, *Assessment of Spawning Horseshoe Crabs;* Monomoy National Wildlife Refuge, *Monomoy National Wildlife Refuge Draft;* Zobel, "Memorandum of Decision."

For community of life supported by horseshoe crabs, see Botton, "The Ecological Importance"; Botton and Shuster, "Horseshoe Crabs in a Food Web" ("the eels ... made a strange sight," 144–45); Buckel and McKown, "Competition"; and Eastern Massachusetts National Wildlife Refuge Complex, *Compatibility Determination*.

For more of the history of Lowcountry plantations, see Coclanis, "Bitter Harvest"; Cuthbert and Hoffius, *Northern Money, Southern Land;* Matthiessen, "Happy Days"; Tufford, *State of Knowledge;* and Tuten, *Lowcountry*.

For protected lands and wetlands along the South Carolina coast, Michael Slattery of the South Carolina Sea Grant Consortium and Coastal Carolina University used GIS mapping data to determine that, of 3,255,000 acres within 20 miles of the coast—still within reach of the tide—902,723 are protected.

For shorebirds in impoundments, see Marsh and Wilkinson, "Significance of the Central Coast of South Carolina as Critical Shorebird Habitat"; Tufford, *State of Knowledge* ("Managed impounded wetlands," 15); and Weber and Haig, "Shorebird Use of South Carolina Managed and Natural Coastal Wetlands."

For conservation funding and excise taxes, see Migratory Bird Conservation Commission, *2012 Annual Report;* President's Task Force, *Final Report* ("Historically, about 90 percent," 7); and U.S. Fish and Wildlife Service, *Budget Justifications and Performance Information, Fiscal Year 2014.*

For erosion in Cape Romain, see Cape Romain National Wildlife Refuge, *Comprehensive Conservation Plan;* and U.S. Fish and Wildlife Service, "Proposed Threatened Status."

For migrating Virginia barrier islands, see Barnes, Truitt, and Warner, *Seashore Chronicles;* and Williams, Dodd, and Gohn, "Coasts in Crisis."

For erosion and sea level rise in Delaware Bay, see Beesley, "Sketch of the Early History"; Delaware Coastal Programs and Delaware Sea Level Rise Advisory Committee, *Preparing for Tomorrow's High Tide;* Dorwart, *Cape May County;* New Jersey Geological Survey, *Geology* ("observations on the dying," 33); Miller et al., "A Geological Perspective"; Murray, "Delaware Gets Millions to Help Beaches"; "Fresh Water Peril"; Niles et al., *Restoration;* Pilkey and Young, *The Rising Sea;* Sweet et al., "Hurricane Sandy"; Tebaldi, Strauss, and Zervas, "Modelling Sea Level Rise"; U.S. Department of the Interior, "Secretary Jewell Announces $102 Million"; U.S. Fish and Wildlife Service, "Proposed Threatened Status"; and U.S. Fish and Wildlife Service, "U.S. Fish and Wildlife to Restore Bay Beaches."

For knots historically in Georgia, see Burleigh, *Georgia Birds;* and Harrington, *The Flight of the Red Knot* ("at least 12,000 knots," 64).

NINE Ghost Trail

For records of wintering knots in Texas, see Morrison and Harrington, "The Migration System."

For piping plover, see U.S. Fish and Wildlife Service, *Piping Plover.*

For sea turtles, see Doughty, "Sea Turtles in Texas"; Hildebrand, "Hallazgo del área"; Neck, "Occurrence of Marine Turtles"; and "Sea Turtle Recovery Project."

For knots in Florida, see Schwarzer et al., "Annual Survival."

For red tide, see Denton and Contreras, *The Red Tide;* Hetland and Campbell, "Convergent Blooms"; Lenes et al., "Saharan Dust"; Magaña, Contreras, and Villareal, "A Historical Assessment" ("foul odor" and "mountain of dead fish," 164); Powell, "Water, Water, Everywhere"; U.S. Fish and Wildlife Service, "Proposed Threatened Status"; and Walsh et al., "Imprudent Fishing" ("the times when the fruit comes to mature and when the fish die," 892).

For redhead ducks, see Woodin and Michot, "Redhead."

For changes in the number of birds and people on Mustang Island, see Foster, Amos, and Fuiman, "Trends in Abundance."

For more on the laguna, see Smith, "Colonial Waterbirds"; Smith, "Redheads"; Tunnell, "The Environment"; and Tunnell, "Geography, Climate, and Hydrography."

More than 2 million birds nesting, wintering, or migrating through the laguna from Bart M. Ballard, Caesar Kleberg Wildlife Research Institute, Texas A&M University.

For wind energy, see American Wind Energy Association, "State Wind Energy Statistics"; Burger et al., "Risk Evaluation"; Chediak, "Gulf Coast Beckons"; de Lucas et al., "Griffon Vulture Mortality"; Manville, "Framing the Issues"; McDonald, "Wind Farms and Deadly Skies"; Smallwood, "Comparing Bird and Bat Fatality-Rate Estimates"; Shawn Smallwood, personal communication, June 30, 2014; Subramanian, "An Ill Wind"; U.S. Department of Energy, *20% Wind Energy;* and Watts, *Wind and Waterbirds* ("buildout of the wind industry along the Atlantic Coast," 1).

For other sources of avian mortality, see Milius, "Cat-Induced Death Toll Revised"; Milius, "Windows Are Major Bird Killers"; and Subramanian, "An Ill Wind."

For wintering and juvenile knots in the Laguna Madre, see Newstead et al., "Geolocation"; U.S. Fish and Wildlife Service, "Rufa Red Knot Ecology and Abundance."

For historical use of the ghost trail, see Cooke, "Distribution and Migration" ("almost endless succession" and "the great highway of spring migration," 5; and "tolerably common," 32); Forbush, *Game Birds* ("diminutive army" and "numbers," 263).

For knots at the prairie pothole lakes, see Alexander et al., "Conventional and Isotopic Determinations"; Alexander and Gratto-Trevor, *Shorebird Migration;* Beyersbergen and Duncan, *Shorebird Abundance;* Newstead et al., "Geolocation"; Niles et al., "Migration Pathways"; Skagen et al., *Biogeographical Profiles;* Thompson, "Record of the Red Knot in Texas"; and WHSRN, "Chaplin Old Wives Reed Lakes."

For the value of the prairie potholes, see Gascoigne et al., "Valuing Ecosystem."

Additional sightings along the central flyway in the United States came from Doug Backland, South Dakota; Joe Grzybowski, for knots in Oklahoma; Lawrence Igl, USGS Northern Prairie Wildlife Research Center in North Dakota, and Dan Svingen, acting district ranger, U.S. Forest Service, North Dakota, who sent "How Lucky Can You Get" from Zimmer, *A Birder's Guide to North Dakota,* 103; and Max Thompson, Kansas.

TEN Does Losing One More Bird Matter?

For woodcock population trends, see Cooper and Rau, *American Woodcock.*

For shorebird declines, see Andres et al., "Population Estimates"; Hicklin and Chardine, "The Morphometrics"; Jehl, "Disappearance"; Morrison et al., "Dramatic Declines"; North American Bird Conservation Initiative Canada, *The State of Canada's Birds, 2012;* Watts and Truitt, "Decline of Whimbrels"; and Zöckler, Lanctot, and Syroechkovsky, "Waders (Shorebirds)."

For "The woodcock is a living refutation," see Leopold, *A Sand County Almanac,* 36.

For the record-breaking flight of the bar-tailed godwit, see Battley et al., "Contrasting Extreme."

For killing of horseshoe crabs in Massachusetts, see Germano, "Horseshoe Crabs."

For sea turtles, see McClenachan, Jackson, and Newman, "Conservation Implications of Historic Sea Turtle Nesting Beach Loss"; Hannan et al., "Dune Vegetation"; Houghton et al., "Jellyfish Aggregations"; King, "Historical Review" ("vessels, which have lost their latitude," 184); Lynam et al., "Jellyfish"; Purcell, Uye, and Lo, "Anthropogenic Causes"; and Wilson et al., *Why Healthy Oceans Need Sea Turtles.*

For honeyguides, see Isack and Reyer, "Honeyguides"; Wheye and Kennedy, *Humans, Nature, and Birds.*

For shorebirds' historical role in eating agricultural pests, see Evenden, "The Laborers of Nature"; and Hornaday, *Our Vanishing Wildlife* ("The protection of shorebirds need not be based," 229; "So great, indeed, is their economic value," 233; "feed upon many of the worst enemies of agriculture," 232; and "among their numerous bird enemies, shorebirds rank high," 229).

For birds as pest control today, see BirdLife International, "Birds Are Very Useful Indicators"; Green and Elmberg, "Ecosystem Services"; Karp et al., "Forest Bolsters Bird Abundance"; and Whelan, Wenny, and Marquis, "Ecosystem Services."

For costs of pesticides, see Hallmann et al., "Declines in Insectivorous Birds"; Hladik, Kolpin, and Kuivila, "Widespread Occurrence of Neonicotinoid Insecticides"; Mineau and Palmer, *The Impact;* Mineau and Whiteside, "Pesticide Acute Toxicity"; Pettis et al., "Crop Pollination"; and Pimentel, "Environmental and Economic Costs."

For birds dispersing seeds, Charlie Crisafulli, U.S. Forest Service, personal communication, September 5, 2014; Dale, Swanson, and Crisafulli, *Ecological Responses to the 1980 Eruption of Mount St. Helens;* Darwin, *On the Origin of Species* ("living birds can hardly fail to be highly," 391); Friðriksson and Magnússon, "Colonization of the Land"; Green and Elmberg, "Ecosystem Services"; Green, Figuerola, and Sánchez, "Implications of Waterbird Ecology"; Kays et al., "The Effect of Feeding Time"; Borgþór Magnússon, personal communication, October 14, 2013; Magnússon, Magnússon, and Friðriksson, "Developments in Plant Colonization"; Nogales et al., "Ecological and Biogeographical Implications"; Sánchez, Green, and Castellanos, "Internal Transport of Seeds"; Şekercioğlu, "Increasing Awareness" ("Perhaps the least appreciated contribution," 465); Wenny et al., "The Need to Quantify"; and Whelan, Wenny, and Marquis, "Ecosystem Services."

For vultures, see Markandya et al., "Counting the Cost of Vulture Decline" ("great service to mankind in keeping clean the environments," 196); Pain et al., "Causes and Effects"; and Wheye and Kennedy, *Humans, Nature, and Birds.*

For Lyme disease, see Blockstein, "Lyme Disease"; Bucher, "The Causes of Extinction"; Ostfeld et al., "Climate, Deer, Rodents, and Acorns"; U.S. Centers for Disease Control and Prevention, "CDC Provides Estimate"; and Zhang et al., "Economic Impact."

For West Nile virus, see Allan et al., "Ecological Correlates"; Kilpatrick, "Globalization"; LaDeau, Kilpatrick, and Marra, "West Nile Virus"; and Swaddle and Calos, "Avian Diversity."

For avian flu, see Altizer, Bartel, and Han, "Animal Migration" ("dense aggregations of animals," 300); Berhane et al., "Highly Pathogenic Avian Influenza"; Brown et al.,

"Dissecting a Wildlife Disease Hotspot"; Brown and Rohani, "The Consequences of Climate Change"; Krauss et al., "Influenza in Migratory Birds"; Krauss et al., "Coincident Ruddy Turnstone Migration"; Maxted et al., "Avian Influenza Virus" ("near-zero prevalence," 329); Maxted et al., "Annual Survival of Ruddy Turnstones"; and David Stallknecht, SCWDS, University of Georgia, personal communication, October 14, 2013 ("Delaware Bay is unique").

For guano, see Mathew, "Peru and the British Guano Market"; Olinger, "The Guano Age in Peru"; and Romero, "Peru Guards Its Guano."

For red-footed boobies, see Galetti and Dirzo, "Ecological and Evolutionary Consequences"; and McCauley et al., "From Wing to Wing."

For extinction rates, see Arkema et al., "Coastal Habitats"; Barnosky et al., "Earth's Sixth Mass Extinction" ("the recent loss," 56); Birdlife International, "One in Eight"; Birdlife International, "We Have Lost Over 150 Bird Species"; Burgess, Bowring, and Shen, "High-Precision Timeline"; Costanza et al., "Changes in the Global Value"; Daily et al., "Ecosystem Services"; Galetti and Dirzo, "Ecological and Evolutionary Consequences"; Green and Elmberg, "Ecosystem Services"; McCauley, "Selling Out on Nature"; Pimm, *The World According to Pimm* ("Humanity's impact," 214); Şekercioğlu, "Increasing Awareness"; Şekercioğlu, Daily, and Ehrlich, "Ecosystem Consequences"; and Zimmer, "The Price Tag."

ELEVEN The Longest Day

For recorded historical observations of Southampton Island, see Comer, "A Geographical Description" ("particularly anxious to make certain inquiries," 87); Manning, "Some Notes"; and Ross, "Whaling."

For population of polar bears in the Foxe Basin, see Peacock et al., "Polar Bear Ecology"; and Stapleton et al., *Aerial Survey.*

For climate of Southampton and East Bay, see CAFF, *Arctic Biodiversity Trends, 2010.*

Additional information about 4KL's travels at www.bandedbirds.org.

For physiological changes in knots on the breeding grounds, see Morrison, Davidson, and Piersma, "Transformations"; and Vézina et al., "Phenotypic Compromises."

For the nineteenth-century history of the search for knots in the Arctic, see Borup, *A Tenderfoot with Peary;* Dresser, "On the Late Dr. Walter's Ornithological Researches"; Ekblaw, "Finding the Nest" ("to ornithologists and bird lovers the world over," 97); Feilden, "Breeding of the Knot"; Feilden, "List of Birds Observed"; Greely, *Three Years of Arctic Service* ("We never obtained the nest" and "a completely-formed hard-shelled egg ready to be laid," 377); Harting, "Discovery of the Eggs"; Hunt and Thompson, *North to the Horizon* ("had never been found," 77); Levere, *Science and the Canadian Arctic;* MacMillan, *How Peary Reached the Pole* ("the eggs had never been found previously," 275); Merriam, "The Eggs of the Knot"; Parmelee, Stephens, and Schmidt, *The Birds of Southeastern Victoria Island* ("most severe, the summit having been swept almost continuously by the polar winds" and "No doubt this individual had," 100); Parry, *Appendix to Captain Parry's Journal* ("killed in the Duke of York's Bay," 355); Parry, *Journal of a Second*

Voyage ("They lay four eggs on a tuft of withered grass, without being at the pains of forming any nest," 460–61; and "zoologist," 344–45); Parry, *Supplement to the Appendix* ("breeds in great abundance on the North Georgian Islands," cci); Pleske, *Birds of the Eurasian Tundra;* and Vaughan, *In Search of Arctic Birds* ("Of all the Arctic breeding waders whose eggs were sought and prized by collectors, the Knot took pride of place" and "hard to explain," 158).

Information on satellite trackers, new light-sensitive data loggers, and geolocators and nesting knots from Paul Howey, Microwave Telemetry, Inc., Columbia, Md.; Niles, "What We Still Don't Know"; Niles et al., "First Results"; and Ron Porter, personal communication, January 2, 2014.

For George Miksch Sutton and other biologists on knots in Southampton, see Abraham and Ankney, "Summer Birds" ("courtship flights, chases, and vocalizations in four separate locations," "broody" female, and "an adult with one flightless young," 184–85); Berger, "George Miksch Sutton"; Jackson, *George Miksch Sutton;* Sutton, "Birds of Southampton Island" ("very quiet and dignified in behavior, especially when compared with the turnstones which flashed and rattled along the beaches everywhere," 123); and Sutton, "The Exploration of Southampton Island" ("clean-edged beauty of the Arctic," 1; "Many an explorer," 1; and "expert mechanic, boatman, huntsman, dog-team driver, and igloo-builder," 5).

For the flight of Arctic terns, see Egevang et al., "Tracking of Arctic Terns."

For knots that "swim with great ease," see Baird, Brewer, and Ridgway, *The Water Birds of North America,* 215.

For preening waxes in sandpipers, see Reneerkens, Piersma, and Damsté, "Sandpipers (Scolopacidae) Switch from Monoester."

Knot nesting in East Bay from Darryl Edwards, Biology Department, Laurentian University, personal communication, February 14, 2014.

For advantages to high-latitude migration, see McKinnon et al., "Lower Predation Risk."

For nest survival when knot parents share the work, see Pirie, Johnston, and Smith, "Tier 2 Surveys."

For lemmings, see Bêty et al., "Shared Predators"; Fraser et al., "The Red Knot"; Nolet et al., "Faltering Lemming Cycles"; Perkins, Smith, and Gilchrist, "The Breeding Ecology of Ruddy Turnstones"; Schmidt et al., "Response of an Arctic Predator."

For breeding knots, see Dresser, "On the Late Dr. Walter's Ornithological Researches" ("the male was always most careful of the young, whereas the female, when in the vicinity, had the appearance of a disinterested spectator," 232); and Parmelee, Stephens, and Schmidt, *The Birds of Southeastern Victoria Island.*

For warming Arctic, see Miller et al., "Unprecedented Recent Summer Warmth"; and Perovich et al., "Sea Ice."

For murres, mosquitoes, and polar bears, see Elliott and Gaston, "Mass-Length Relationships"; Gaston and Elliott, "Effects of Climate-Induced Changes"; Gaston, Smith, and Provencher, "Discontinuous Change"; Tony Gaston, research scientist (ret.), Environment Canada, personal communication, February 4, 2014 ("so thick," "they are wearing fur boots," and "It just won't be the Arctic"); Mallory et al., "Effects of Climate Change"; Smith et al., "Has Early Ice Clearance Increased Predation?"; Zöckler, Lanctot, and Syroechkovsky, "Waders (Shorebirds)."

For melting ice and polar bears in Hudson Bay, see Castro de la Guardia et al., "Future Sea Ice Conditions"; Iverson et al., "Longer Ice-Free Seasons"; Molnár et al., "Predicting Climate Change Impacts"; Molnár et al., "Predicting Survival"; Karyn Rode, U.S. Geological Survey, personal communication, September 8, 2014; Rode et al., "Comments in Response"; Rode et al., "Variation in the Response"; and Stirling and Derocher, "Effects of Climate Warming."

For geese, see Abraham et al., "Northern Wetland Ecosystems"; Alisauskas, Leafloor, and Kellet, "Population Status"; AMAP, *Arctic Climate Issues 2011*; Feng et al., "Evaluating"; Johnson et al., "Assessment of Harvest"; Kerbes, Meeres, and Alisaukas, *Surveys of Nesting Lesser Snow Geese*; Jim Leafloor, Canadian Wildlife Service, personal communication, January 28, 2014 ("about 29 polar bears"); Paul Smith, Environment Canada, personal communication, January 23, 2014 ("tanking" and "thick"); and Smith et al., "Has Early Ice Clearance Increased Predation?"

Almost finding a knot nest in East Bay from Kara Anne Ward, medical student, University of Ottawa, personal communication, August 22, 2012.

TWELVE Returning South

For numbers of shorebirds and knots going through James Bay, see Mark Peck, Royal Ontario Museum, personal communication, April 2, 2014; and Pollock, Abraham, and Nol, "Migrant Shorebird Use of Akimiski Island."

For possible first knots on James Bay, see Newman, *A Dictionary of British Birds;* Richardson, Swainson, and Kirby, *Fauna Boreali-Americana;* and Williams, *Andrew Graham's Observations.*

Sewage lagoon counts provided by Mike Burrell, Bird Studies Canada.

Bird counts at Longridge provided by Jean Iron, Ontario.

For more recent observations of knots passing through the bay and the food they eat, see Hope and Shortt, "Southward Migration of Adult Shorebirds" ("an enormous flock," 572); Martini and Morrison, "Regional Distribution"; Morrison and Harrington, "Critical Shorebird Resources"; and Morrison and Harrington, "The Migration System of the Red Knot."

For the Hudson Bay Lowland, see Abraham and Keddy, "The Hudson Bay Lowland"; Abraham et al., "Hudson Plains Ecozone+ Status and Trends Assessment"; Riley, *Wetlands of the Ontario Hudson Bay Lowland;* and Stewart and Lockhart, *An Overview of the Hudson Bay Marine Ecosystem.*

Chickney Point data from Christian Friis, Canadian Wildlife Service, and the team at Chickney Point.

For James Bay shrinking at Shegogau Creek, Ken Abraham, Ontario Ministry of Natural Resources, personal communication, February 27, 2014.

Possible origin of fossil from David Rudkin, Royal Ontario Museum, personal communication, October 3, 2012.

Jean Iron's website is http://www.jeaniron.ca/.

For starlings, see Cabe, "European Starling (*Sturnus Vulgaris*)" ("arguably the most successful avian introduction on this continent").

For the Bonaparte's gull, see Burger and Gochfeld, "Bonaparte's Gull."

For nesting little gulls, see Wilson and McRae, *Seasonal and Geographical Distribution of Birds.*

For 1994 knot sightings in Churchill, see IBA Canada, "Churchill and Vicinity."

For geolocators and Nelson River, see Niles et al., "Migration Pathways"; and Niles et al., "First Results."

Information on the new radio sensors from Phil Taylor, Bird Studies Canada Chair of Ornithology at Acadia University, Nova Scotia, personal communication, February 21, 2014.

Information for knots along the Hayes River from Ann McKellar, Canadian Wildlife Service, personal communication, June 26, 2014.

Bird sightings from the James Bay team on the way back to Toronto from Mike Burrell.

Big Year birds finding from Josh Vandermeulen, personal communication, February 11, 2014, account on http://joshvandermeulen.blogspot.com/.

Knot data on the Mingan Islands comes from Yves Aubry, Canadian Wildlife Service.

For ocean acidification, see Benoît et al., *State-of-the-Ocean Report;* Cooley and Doney, "Anticipating Ocean Acidification's Economic Consequences"; Gaylord et al., "Functional Impacts of Ocean Acidification"; Gobler et al., "Hypoxia and Acidification"; Hönisch et al., "The Geological Record of Ocean Acidification"; IGBP, IOC, and SCOR, *IGBP, IOC, SCOR: Ocean Acidification Summary for Policymakers;* Jansson, Norkko, and Norkko, "Effects of Reduced pH on *Macoma Balthica*"; O'Donnell, George, and Carrington, "Mussel Byssus Attachment"; Talmage and Gobler, "Effects of Past, Present, and Future Ocean Carbon Dioxide"; Van Colen et al., "The Early Life History of the Clam *Macoma Balthica*"; Waldbusser et al., "A Developmental and Energetic Basis"; Waldbusser and Salisbury, "Ocean Acidification in the Coastal Zone"; and Wang et al., "The Marine Inorganic Carbon System."

For shorebirds in the Guianas, see Morrison and Ross, *Atlas of Nearctic Shorebirds on the Coast of South America*, vol. 1; Morrison and Spaans, "National Geographic Mini-Expedition to Surinam, 1978"; Morrison et al., "Dramatic Declines"; Ottema and Spaans, "Challenges and Advances in Shorebird Conservation"; Trull, "Shorebirds and Noodles" ("strongly-tasting," 269); U.S. Fish and Wildlife Service, "Proposed Threatened Status"; U.S. Fish and Wildlife Service, "Rufa Red Knot Ecology and Abundance"; and Watts et al., "Whimbrel Tracking in the Americas."

EPILOGUE

For seals and sealing, see Lelli, Harris, and Aboueissa, "Seal Bounties in Maine and Massachusetts"; Mowat, *Sea of Slaughter.*

For history of knots and horseshoe crabs on Cape Cod, see Forbush, *Game Birds;*
   Hapgood and Roosevelt, *Shorebirds;* Mackay, "Observations on the Knot"
   ("exceedingly large numbers" and "This was previous," 29; and "in old times," 30);
   and Thoreau, *Cape Cod.*
For wintering knots molting in Massachusetts, see Burger et al., "Migration and Over-
   wintering of Red Knots"; Niles et al., "Migration Pathways."
For juvenile survival, see Leyrer et al., "Small-Scale Demographic Structure."
For knots historically near my home, see Townsend, *Birds of Essex County.*
For knots and hurricanes, see Niles et al., "Migration Pathways"; and Niles et al., "First
   Results."

Bibliography

Abraham, K. F., and C. D. Ankney. "Summer Birds of East Bay, Southampton Island, Northwest Territories." *Canadian Field-Naturalist* 100, no. 2 (1986): 180–85.

Abraham, K. F., R. L. Jefferies, R. T. Alisaukas, and R. F. Rockwell. "Northern Wetland Ecosystems and Their Response to High Densities of Lesser Snow Geese and Ross's Geese." In *Evaluation of Special Management Measures for Midcontinent Lesser Snow Geese and Ross's Geese*, edited by J. O. Leafloor, T. J. Moser, and B. D. J. Batt, 9–45. Arctic Goose Joint Venture Special Publication. Washington, D.C., and Ottawa: U.S. Fish and Wildlife Service and Canadian Wildlife Service, 2012.

Abraham, K. F., and C. J. Keddy. "The Hudson Bay Lowland." In *The World's Largest Wetlands*, edited by Lauchlan H. Fraser and Paul A. Keddy, 118–48. Cambridge: Cambridge University Press, 2005.

Abraham, K. F., L. M. McKinnon, Z. Jumean, S. M. Tully, L. R. Walton, and H. M. Stewart. "Hudson Plains Ecozone[+] Status and Trends Assessment." Ottawa: Canadian Council of Resource Ministers, 2011.

Aieta, Amy E., and Kenneth Oliveira. "Distribution, Prevalence, and Intensity of the Swim Bladder Parasite *Anguillicola Crassus* in New England and Eastern Canada." *Diseases of Aquatic Organisms* 84, no. 3 (2009): 229–35.

Albrieu, C., and S. Ferrari. "La participación de los municipios en la conservación de los humedales costeros." Presentation at the Taller Regional sobre Humedales Costeros Patagónicos. Organizado por la Secretaría de Ambiente y Desarrollo Sustentable de la Nación, Buenos Aires, July 2–3, 2007.

Albrieu, C., S. Ferrari, and G. Montero. "Investigación, educación e transferencia: Unha alianza para a conservación das aves de praia migratorias e os seus hábitats no estuario do Río Gallegos (Patagonia Austral, Argentina)." *AmbientaMENTE Sustentable* 1, nos. 9–10 (2010): 18–97.

Alexander, Stuart, and Cheri L. Gratto-Trevor. *Shorebird Migration and Staging at a Large Prairie Lake and Wetland Complex: The Quill Lakes, Saskatchewan.* Canadian Wildlife Service, 1997.

Alexander, S. A., K. A. Hobson, C. L. Gratto-Trevor, and A. W. Diamond. "Conventional and Isotopic Determinations of Shorebird Diets at an Inland Stopover: The Importance of Inverterbrates and *Potamogeton Pectinatus* Tubers." *Canadian Journal of Zoology* 74, no. 6 (1996): 1057–68.

Alisauskas, R. T., J. O. Leafloor, and D. K. Kellet. "Population Status of Midcontinent Lesser Snow Geese and Ross's Geese Following Special Conservation Measures." In *Evaluation of Special Management Measures for Midcontinent Lesser Snow Geese and Ross's Geese.*, edited by J. O. Leafloor, T. J. Moser, and B. D. J. Batt, 132–77. Arctic Goose Joint Venture Special Publication. Washington, D.C., and Ottawa: U.S. Fish and Wildlife Service and Canadian Wildlife Service, 2012.

Allan, Brian F., R. Brian Langerhans, Wade A. Ryberg, William J. Landesman, Nicholas W. Griffin, Rachael S. Katz, Brad J. Oberle, et al. "Ecological Correlates of Risk and Incidence of West Nile Virus in the United States." *Oecologia* 158, no. 4 (2009): 699–708.

Altizer, Sonia, Rebecca Bartel, and Barbara A. Han. "Animal Migration and Infectious Disease Risk." *Science* 331, no. 6015 (2011): 296–302.

AMAP. *Arctic Climate Issues 2011: Changes in Arctic Snow, Water, Ice, and Permafrost. SWIPA 2011 Overview Report.* Oslo: AMAP, 2012.

American Wind Energy Association. "State Wind Energy Statistics: Texas," June 3, 2013. http://www.awea.org/Resources/state.aspx?ItemNumber=5183.

Anderson, Rebecca L., Winsor H. Watson, and Christopher C. Chabot. "Sublethal Behavioral and Physiological Effects of the Biomedical Bleeding Process on the American Horseshoe Crab, *Limulus Polyphemus*." *Biological Bulletin*, December 1, 2013, 137–51.

Andres, Brad A., Paul A. Smith, R. I. Guy Morrison, Cheri L. Gratto-Trevor, Stephen C. Brown, and Christian A. Friis. "Population Estimates of North American Shorebirds, 2012." *Wader Study Group Bulletin* 119 (2013): 178–94.

The Aransas Project v Shaw, et al. No. 13-40317, U.S. Court of Appeals, Fifth Circuit, 2014.

Arkema, Katie K., Greg Guannel, Gregory Verutes, Spencer A. Wood, Anne Guerry, Mary Ruckelshaus, Peter Kareiva, Martin Lacayo, and Jessica M. Silver. "Coastal Habitats Shield People and Property from Sea-Level Rise and Storms." *Nature Climate Change* 3, no. 10 (2013): 913–18.

Atkinson, Philip W., Allan J. Baker, Karen A. Bennett, Nigel A. Clark, Jacquie A. Clark, Kimberly B. Cole, Anne Dekinga, Amanda Dey, Simon Gillings, and Patricia M. González. "Rates of Mass Gain and Energy Deposition in Red Knot on Their Final Spring Staging Site Is Both Time- and Condition-Dependent." *Journal of Applied Ecology* 44, no. 4 (2007): 885–95.

Atlantic States Marine Fisheries Commission (ASMFC). *Addendum I to the Fishery Management Plan for Horseshoe Crab.* Arlington, Va.: ASMFC, April 2000.

———. *American Eel Benchmark Stock Assessment.* Stock Assessment Report no. 12–01. Arlington, Va.: ASMFC, 2012.

———. *American Shad Stock Assessment.* Report no. 07–01 (supplement), vol. 51. Arlington, Va.: ASMFC, 2007.

———. *ASMFC Approves Resolution to Ban the Import and Use of Asian Horseshoe Crabs as Bait.* Arlington, Va.: ASMFC, February 21, 2013.

———. *ASMFC River Herring Benchmark Assessment Indicates Stock Is Depleted.* Arlington, Va.: ASMFC, May 4, 2012.

———. *Draft Addendum III to the Fishery Management Plan for American Eel for Public Comment.* Arlington, Va.: ASMFC, March 2013.

———. *Horseshoe Crab Technical Committee Meeting Summary,* September 25, 2013. Arlington, Va.: ASMFC, n.d.

Atlantic States Marine Fisheries Commission (ASMFC) Delaware Bay Ecosystem Technical Committee. *ARM Recommendation.* Arlington, Va.: ASMFC, September 5, 2012.

———. *Meeting Summary,* September 24, 2013. Arlington, Va.: ASMFC, n.d.

Atlantic States Marine Fisheries Commission (ASMFC) Horseshoe Crab Plan Review Team. *2013 Review of the Atlantic States Marine Fisheries Commission Fishery Management Plan for Horseshoe Crab (Limulus Polyphemus) 2012 Fishing Year.* Arlington, Va.: ASMFC, May 2013.

Atlantic States Marine Fisheries Commission (ASMFC) Horseshoe Crab Stock Assessment Subcommittee. *2013 Horseshoe Crab Stock Assessment Update.* Arlington, Va.: ASMFC, 2013.

Audubon, John James, and William Macgillivray. *Ornithological Biography.* Vol. 1. Pittsburgh: University of Pittsburgh, 2007. http://digital.library.pitt.edu/cgi-bin/t/text/text-idx?idno=31735056284882;view=toc;c=darltext.

———. *Ornithological Biography.* Vol. 3. Philadelphia: Judah Dobson, A. Black, 1839. http://digital.library.pitt.edu/cgi-bin/t/text/text-idx?c=darltext&cc=darltext&type=simple&q1=ornithological+biography&button1=Go.

———. *Ornithological Biography.* Vol. 4. Philadelphia: Judah Dobson, A. Black, 1839. http://digital.library.pitt.edu/cgi-bin/t/text/text-idx?c=darltext&cc=darltext&type=simple&q1=ornithological+biography&button1=Go.

Baird, Spencer Fullerton, T. M. Brewer, and Robert Ridgway. *The Water Birds of North America.* Boston: Little, Brown, 1884. http://archive.org/details/waterbirdsofnort02bair.

Baker, Allan J., Patricia M. González, Theunis Piersma, Lawrence J. Niles, Ines de Lima Serrano do Nascimento, Philip W. Atkinson, Nigel A. Clark, Clive D. T. Minton, Mark K. Peck, and Geert Aarts. "Rapid Population Decline in Red Knots: Fitness Consequences of Decreased Refuelling Rates and Late Arrival in Delaware Bay." *Proceedings of the Royal Society of London, Series B: Biological Sciences* 271, no. 1541 (2004): 875–82.

Baker, Allan J., Sergio L. Pereira, and Tara A. Paton. "Phylogenetic Relationships and Divergence Times of Charadriiformes Genera: Multigene Evidence for the Cretaceous Origin of at Least 14 Clades of Shorebirds." *Biology Letters*, April 22, 2007, 205–9.

Baker, Allan J., Theunis Piersma, and Lene Rosenmeier. "Unraveling the Intraspecific Phylogeography of Knots *Calidris Canutus*: A Progress Report on the Search for Genetic Markers." *Journal für Ornithologie*, October 1, 1994, 599–608.

Baldi, Ricardo, Andrés Novaro, Martín Funes, Susan Walker, Pablo Ferrando, Mauricio Failla, and Pablo Carmanchahi. "Guanaco Management in Patagonian Rangelands: A Conservation Opportunity on the Brink of Collapse." In *Wild Rangelands: Conserving Wildlife While Maintaining Livestock in Semi-arid Ecosystems*, edited by Johan T. du Toit Head, Richard Kocknager, and James C. Deutsch, 266–90. Hoboken: John Wiley & Sons, 2010.

Ball, Walter William Rouse. *A History of the Study of Mathematics at Cambridge*. Cambridge: Cambridge University Press, 1883.

Bandedbirds.org. "Banding and Resightings." www.bandedbirds.org.

Banerji, Mary Ann, and Richard P. Spencer. "Febrile Response to Cerebrospinal Fluid Flow Studies." *Journal of Nuclear Medicine* 13, no. 8 (1972): 655.

Barlow, Robert B., and Maureen K. Powers. "Seeing at Night and Finding Mates: The Role of Vision." In *The American Horseshoe Crab*, edited by Carl N. Shuster Jr., Robert B. Barlow, and H. Jane Brockmann, 83–102. Cambridge, Mass.: Harvard University Press, 2003.

Barnes, Brooks, Barry R. Truitt, and William A. Warner. *Seashore Chronicles*. Charlottesville: University of Virginia, 1999.

Barnosky, Anthony D., and Emily L. Lindsey. "Timing of Quaternary Megafaunal Extinction in South America in Relation to Human Arrival and Climate Change." *Quaternary International*, April 15, 2010, 10–29.

Barnosky, Anthony D., Nicholas Matzke, Susumu Tomiya, Guinevere O. U. Wogan, Brian Swartz, Tiago B. Quental, Charles Marshall, et al. "Has the Earth's Sixth Mass Extinction Already Arrived?" *Nature* 471, no. 7336 (2011): 51–57.

Barsoum, Noha, and Charles Kleeman. "Now and Then, the History of Parenteral Fluid Administration." *American Journal of Nephrology* 22, nos. 2–3 (2002): 284–89.

Bastidas, Pamela Domínquez. *Cerro Sombrero: Arquitectura moderna en Tierra del Fuego*. Santiago: CNCA, 2011.

Battley, Phil F., Nils Warnock, T. Lee Tibbitts, Robert E. Gill, Theunis Piersma, Chris J. Hassell, David C. Douglas, et al. "Contrasting Extreme Long-Distance Migration Patterns in Bar-Tailed Godwits *Limosa Lapponica*." *Journal of Avian Biology* 43, no. 1 (2012): 21–32.

Beesley, Maurice, M.D. "Sketch of the Early History of the County of Cape May." In New Jersey Geological Survey, *Geology of the County of Cape May, State of New Jersey*, 158–205. Trenton: Printed at the Office of the True American, 1857.

Bell, J. C., and J. T. Nichols. "Notes on the Food of Carolina Sharks." *Copeia*, March 15, 1921, 17–20.

Benoît, Hugues P., Jacques A. Gagné, Patrick Ouellet, and Marie-Noëlle Bourassa, eds. *State-of-the-Ocean Report for the Gulf of St. Lawrence Integrated Management (GOSLIM) Area*. Moncton, New Brunswick: Fisheries and Oceans Canada; Mont-Joli, Québec: Pêche et Océans, 2012.

Bent, Arthur Cleveland. *Life Histories of North American Shore Birds*. Vol. 1. New York: Dover, 1962.

"Berge Nice." *Ocean Orbit*, February 2005, 3.

Berger, Andrew J. "George Miksch Sutton." *Wilson Bulletin* 80, no. 1 (1968): 30–35.

Bergreen, Laurence. *Over the Edge of the World*. New York: William Morrow, 2003.

Berhane, Yohannes, Tamiko Hisanaga, Helen Kehler, James Neufeld, Lisa Manning, Connie Argue, Katherine Handel, Kathleen Hooper-McGrevy, Marilyn Jonas, and John Robinson. "Highly Pathogenic Avian Influenza Virus A (H7N3) in Domestic Poultry, Saskatchewan, Canada, 2007." *Emerging Infectious Diseases* 15, no. 9 (2009): 1492.

Bêty, Joël, Gilles Gauthier, Erkki Korpimäki, and Jean-François Giroux. "Shared Predators and Indirect Trophic Interactions: Lemming Cycles and Arctic-Nesting Geese." *Journal of Animal Ecology* 71, no. 1 (2002): 88–98.

Beyersbergen, Gerard W., and David C. Duncan. *Shorebird Abundance and Migration Chronology at Chaplin Lake, Old Wives Lake and Reed Lake, Saskatchewan: 1993 and 1994*. Technical Report Series no. 484. Environment Canada, 2007.

Bird, Junius. *Travels and Archaeology in South Chile*. Edited by John Hyslop. Iowa City: University of Iowa Press, 1988.

BirdLife International. "Birds Are Very Useful Indicators for Other Kinds of Biodiversity." *Birdlife International: State of the World's Birds*, 2013. http://www.birdlife.org/datazone/sowb/casestudy/79.

———. "*Eurynorhynchus Pygmeus*." *IUCN 2013: IUCN Red List of Threatened Species, Version 2013.2.*, 2013. www.iucnredlist.org.

———. "One in Eight of All Bird Species Is Threatened with Global Extinction." *Birdlife International: State of the World's Birds*, 2014. http://www.birdlife.org/datazone/sowb/casestudy/106.

———. "*Podiceps Gallardoi*." *IUCN Red List of Threatened Species Version 2013.1*, 2013. http://www.iucnredlist.org.

———. "We Have Lost Over 150 Bird Species since 1500." *Birdlife International: State of the World's Birds*, 2011. http://www.birdlife.org/datazone/sowb/casestudy/102.

Bjorndal, Karen A., and Jeremy B. C. Jackson. "Roles of Sea Turtles in Marine Ecosystems: Reconstructing the Past." In *The Biology of Sea Turtles*, edited by Peter L. Lutz, John A. Musick, and Jeanette Wyneken, 2:259–74. Boca Raton: CRC, 2003.

Blockstein, David E. "Lyme Disease and the Passenger Pigeon?" *Science*, March 20, 1998, 1831.

Borup, George. *A Tenderfoot with Peary*. New York: F. A. Stokes, 1911.

Botton, Mark L. "The Ecological Importance of Horseshoe Crabs in Estuarine and Coastal Communities: A Review and Speculative Summary." In *Biology and*

Conservation of Horseshoe Crabs, edited by John T. Tanacredi, Mark L. Botton, and David R. Smith, 45–63. Dordrecht: Springer, 2009.

Botton, M. L., and B. A. Harrington. "Synchronies in Migration: Shorebirds, Horseshoe Crabs, and Delaware Bay." In *The American Horseshoe Crab*, edited by Carl N. Shuster Jr., Robert B. Barlow, and H. Jane Brockmann, 5–26. Cambridge, Mass.: Harvard University Press, 2003.

Botton, Mark L., and Tomio Ito. "The Effects of Water Quality on Horseshoe Crab Embryos and Larvae." In *Biology and Conservation of Horseshoe Crabs*, edited by John T. Tanacredi, Mark L. Botton, and David R. Smith, 439–52. Dordrecht: Springer, 2009.

Botton, M. L., and R. E. Loveland. "Reproductive Risk: High Mortality Associated with Spawning by Horseshoe Crabs (*Limulus Polyphemus*) in Delaware Bay, USA." *Marine Biology*, April 1, 1989, 143–51.

Botton, Mark L., Robert E. Loveland, and Timothy R. Jacobsen. "Site Selection by Migratory Shorebirds in Delaware Bay, and Its Relationship to Beach Characteristics and Abundance of Horseshoe Crab (*Limulus Polyphemus*) Eggs." *Auk* 111, no. 3 (1994): 605–16.

Botton, M., P. Shin, S. Cheung, G. Gauvry, G. Kreamer, D. Smith, J. Tanacredi, and K. Laurie. "Emerging Issues in Horseshoe Crab Conservation: A Perspective from the IUCN Species Specialist Group." In *Abstract Book*, 21. San Diego: CERF, 2013.

Botton, M. L., and Carl N. Shuster Jr. "Horseshoe Crabs in a Food Web: Who Eats Whom?" In *The American Horseshoe Crab*, edited by Carl N. Shuster Jr., Robert B. Barlow, and H. Jane Brockmann, 133–53. Cambridge, Mass.: Harvard University Press, 2003.

Brown, V. L., J. M. Drake, D. E. Stallknecht, J. D. Brown, K. Pedersen, and P. Rohani. "Dissecting a Wildlife Disease Hotspot: The Impact of Multiple Host Species, Environmental Transmission and Seasonality in Migration, Breeding and Mortality." *Journal of the Royal Society Interface* 10, no. 79 (2013): 20120804.

Brown, V. L., and Pejman Rohani. "The Consequences of Climate Change at an Avian Influenza 'Hotspot.'" *Biology Letters* 8, no. 6 (2012): 1036–39.

Bucher, Enrique H. "The Causes of Extinction of the Passenger Pigeon." *Current Ornithology* 9 (1992): 1–36.

Buckel, Jeffrey A., and Kim A. McKown. "Competition between Juvenile Striped Bass and Bluefish: Resource Partitioning and Growth Rate." *Marine Ecology Progress Series* 234 (2002): 191–204.

Buehler, Deborah M., and Allan J. Baker. "Population Divergence Times and Historical Demography in Red Knots and Dunlins." *Condor* 107, no. 3 (2005): 497–513.

Buehler, Deborah M., Allan J. Baker, and Theunis Piersma. "Reconstructing Palaeoflyways of the Late Pleistocene and Early Holocene Red Knot *Calidris Canutus*." *Ardea* 94, no. 3 (2006): 484–98.

Burger, Joanna, and Michael Gochfeld. "Bonaparte's Gull (*Larus Philadelphia*)." Edited by A. Poole and F. Gill. *The Birds of North America Online*, 2002.

Burger, Joanna, Caleb Gordon, J. Lawrence, James Newman, Greg Forcey, and Lucy Vlietstra. "Risk Evaluation for Federally Listed (Roseate Tern, Piping Plover) or Candidate (Red Knot) Bird Species in Offshore Waters: A First Step for Managing the Potential Impacts of Wind Facility Development on the Atlantic Outer Continental Shelf." *Renewable Energy* 36, no. 1 (2011): 338–51.

Burger, Joanna, and Lawrence J. Niles. "Closure versus Voluntary Avoidance as a Method of Protecting Migrating Shorebirds on Beaches in New Jersey." *Wader Study Group Bulletin* 120, no. 1 (2013): 20–25.

Burger, Joanna, Lawrence J. Niles, Ronald R. Porter, Amanda D. Dey, Stephanie Koch, and Caleb Gorden. "Migration and Over-wintering of Red Knots (*Calidris Canutus Rufa*) along the Atlantic Coast of the United States." *Condor* 114, no. 2 (2012): 1–12.

Burgess, Seth D., Samuel Bowring, and Shu-zhong Shen. "High-Precision Timeline for Earth's Most Severe Extinction." *Proceedings of the National Academy of Sciences* 111, no. 9 (2014): 3316–21.

Burleigh, Thomas. *Georgia Birds*. Norman: University of Oklahoma Press, 1958.

Cabe, Paul R. "European Starling (*Sturnus Vulgaris*)." Edited by A. Poole and F. Gill. *The Birds of North America Online*, 1993.

CAFF. *Arctic Biodiversity Trends, 2010—Selected Indicators of Change*. Akureyri, Iceland: CAFF International Secretariat, May 2010.

Cantwell, Robert. *Alexander Wilson: Naturalist and Pioneer, a Biography*. Philadelphia: Lippincott, 1961.

Cape Romain National Wildlife Refuge. *Comprehensive Conservation Plan*. Atlanta: USFWS Southeast Region, 2010.

Carmona, Roberto, Victor Ayala-Pérez, Nallely Arce, and Lorena Morales-Gopar. "Use of Saltworks by Red Knots at Guerrero Negro, Mexico." *Wader Study Group Bulletin* 111 (2006): 46–49.

Carpenter, C. C. "The Erratic Evolution of Cholera Therapy: From Folklore to Science." *Clinical Therapeutics* 12, supplement A (1990): 22–28.

Castro de la Guardia, Laura, Andrew E. Derocher, Paul G. Myers, Arjen D. Terwisscha van Scheltinga, and Nick J. Lunn. "Future Sea Ice Conditions in Western Hudson Bay and Consequences for Polar Bears in the 21st Century." *Global Change Biology* 19, no. 9 (2013): 2675–87.

Chediak, Mark. "Gulf Coast Beckons Wind Farms When West Texas Gusts Fade." *Bloomberg Sustainability*, October 11, 2013. http://www.bloomberg.com/news/2013–10–10/gulf-coast-beckons-wind-farms-when-west-texas-gusts-fade.html.

Chen, Chang-Po, and Hwey-Lian Hsieh. "The Challenges and Opportunities for Horseshoe Crab Conservation in Taiwan." Presentation at the International Workshop on the Science and Conservation of Asian Horseshoe Crabs, Hong Kong, June 12–16, 2011. In *Abstracts of Plenary Talks and Oral Presentations*, PT-1. Hong Kong, 2011. http://www.cityu.edu.hk/bch/iwscahc2011/Download/Plenary_Talks_&_Oral.pdf.

"The Cholera." *Lancet* 2 (1854): 266.

Cione, Alberto L., Eduardo P. Tonni, and Leopoldo Soibelzon. "Did Humans Cause the Late Pleistocene–Early Holocene Mammalian Extinctions in South America in a Context of Shrinking Open Areas?" In *American Megafaunal Extinctions at the End of the Pleistocene*, edited by Gary Haynes, 125–44. Dordrecht: Springer Netherlands, 2009.

Clark, Kathleen E., Lawrence J. Niles, and Joanna Burger. "Abundance and Distribution of Migrant Shorebirds in Delaware Bay." *Condor* 95, no. 3 (1993): 694–705.

Cobb, J. N. "The Sturgeon Fishery of Delaware River and Bay." In *U.S. Fish Commission Report for 1899*, 369–80. Washington, D.C.: U.S. Commission of Fish and Fisheries, 1900.

Coclanis, Peter A. "Bitter Harvest: The South Carolina Low Country in Historical Perspective." *Journal of Economic History* 45, no. 2 (1985): 251–59.

Cohen, Jonathan B., Brian D. Gerber, Sarah M. Karpanty, James D. Fraser, and Barry R. Truitt. "Day and Night Foraging of Red Knots (*Calidris Canutus*) during Spring Stopover in Virginia, USA." *Waterbirds* 34, no. 3 (2011): 352–56.

Comer, George. "A Geographical Description of Southampton Island and Notes upon the Eskimo." *Bulletin of the American Geographical Society* 42, no. 2 (1910): 84–90.

Cooke, Wells. "Distribution and Migration of North American Shorebirds." *Bulletin of the United States Bureau of Biological Survey* 35 (1912). http://www.biodiversitylibrary.org/bibliography/54050.

Cooley, Sarah R., and Scott C. Doney. "Anticipating Ocean Acidification's Economic Consequences for Commercial Fisheries." *Environmental Research Letters*, June 1, 2009, 024007.

Cooper, James F., and John C. Harbert. "Endotoxin as a Cause of Aseptic Meningitis after Radionuclide Cisternography." *Journal of Nuclear Medicine* 16, no. 9 (1975): 809–13.

Cooper, T. R., and R. D. Rau. *American Woodcock Population Status, 2013.* Laurel, Md.: U.S. Fish and Wildlife Service, 2013.

Cornell Lab of Ornithology. "Eskimo Curlew: Three Strikes in the Wink of an Eye." *All about Birds*, n.d. www.birds.cornell.edu/AllAbout Birds/conservation/extinctions/eskimo-curlew.

Costanza, Robert, Rudolf de Groot, Paul Sutton, Sander van der Ploeg, Sharolyn J. Anderson, Ida Kubiszewski, Stephen Farber, and R. Kerry Turner. "Changes in the Global Value of Ecosystem Services." *Global Environmental Change* 26 (May 2014): 152–58.

Cuthbert, Robert B., and Stephen G. Hoffius, eds. *Northern Money, Southern Land.* Columbia: University of South Carolina Press, 2009.

Daily, Gretchen C., Stephen Polasky, Joshua Goldstein, Peter M. Kareiva, Harold A. Mooney, Liba Pejchar, Taylor H. Ricketts, James Salzman, and Robert Shallenberger. "Ecosystem Services in Decision Making: Time to Deliver." *Frontiers in Ecology and the Environment* 7, no. 1 (2009): 21–28.

Dale, Virginia H., Frederick J. Swanson, and Charles M. Crisafulli. *Ecological Responses to the 1980 Eruption of Mount St. Helens.* New York: Springer, 2005.

Darwin, Charles. *The Autobiography of Charles Darwin, 1809–1882. With the Original Omissions Restored. Edited and with Appendix and Notes by His Grand-daughter Nora Barlow*. London: Collins, 1958.

———. *Journal of Researches into the Natural History and Geology of the Countries Visited during the Voyage of H.M.S.* Beagle *round the World*. 2nd ed. London: John Murray, 1845. http://darwin-online.org.uk/content/frameset?itemID=F14 &viewtype=text&pageseq=1.

———. *On the Origin of Species by Means of Natural Selection; or, The Preservation of Favoured Races in the Struggle for Life*. 3rd ed. London: John Murray, 1861. http://darwin-online.org.uk/content/frameset?itemID=F381&viewtype=text& pageseq=1.

Davis, Michelle L., Jim Berkson, and Marcella Kelly. "A Production Modeling Approach to the Assessment of the Horseshoe Crab (*Limulus Polyphemus*) Population in Delaware Bay." *Fishery Bulletin* 104 (2006): 215–26.

Delancey, Larry, and Brad Floyd. *Tagging of Horseshoe Crabs*, Limulus Polyphemus, *in Conjunction with Commercial Harvesters and the Biomedical Industry in South Carolina*. Charleston: South Carolina Sea Grant Consortium, 2012.

Delaware Coastal Programs and Delaware Sea Level Rise Advisory Committee. *Preparing for Tomorrow's High Tide*. Dover: Delaware Coastal Programs, Department of Natural Resources and Environmental Control, July 2012.

De Lucas, Manuela, Miguel Ferrer, Marc J. Bechard, and Antonio R. Muñoz. "Griffon Vulture Mortality at Wind Farms in Southern Spain: Distribution of Fatalities and Active Mitigation Measures." *Biological Conservation* 147, no. 1 (2012): 184–89.

Denton, Winston, and Cindy Contreras. *The Red Tide (*Karenia Brevis*) Bloom of 2000*. Austin: Resource Protection Division, Texas Parks and Wildlife Department, June 2004.

DeVillers, Pierre, and Jean A. Terschuren. "Some Distributional Records of Migrant North American Charadriiformes in Coastal South America." *Le Gerfaut* 67 (1977): 107–25.

Dey, Amanda, Kevin Kalasz, and Dan Hernandez. "Delaware Bay Egg Survey, 2005–2010: Unpublished Report to the Atlantic States Marine Fisheries Commission," 2011.

Dey, Amanda, Matthew Danihel, Kevin Kalasz, and Dan Hernandez. "Delaware Bay Horseshoe Crab Egg Survey, 2005–2012: Unpublished Report to the Atlantic States Marine Fisheries Commission," 2013.

Ding, Jeak Ling, and Bow Ho. "Endotoxin Detection—From Limulus Amebocyte Lysate to Recombinant Factor C." In *Endotoxins: Structure, Function and Recognition*, edited by Xiaoyuan Wang and Peter J. Quinn, 187–208. Subcellular Biochemistry 53. Springer 2010. http://link.springer.com/chapter/10.1007/978–90–481–9078–2_9.

———. "Strategy to Conserve Horseshoe Crabs by Genetic Engineering of Limulus Factor C for Pyrogen Testing." Presentation at the International Workshop on the Science and Conservation of Asian Horseshoe Crabs, Hong Kong,

June 12–16, 2011. In *Abstracts of Plenary Talks and Oral Presentations*, O–11. Hong Kong, 2011.

Ding, Jeak Ling, Yong Zhu, and Bow Ho. "High-Performance Affinity Capture-Removal of Bacterial Pyrogen from Solutions." *Journal of Chromatography B: Biomedical Sciences and Applications* 759, no. 2 (2001): 237–46.

Dorwart, Jeffery M. *Cape May County, New Jersey: The Making of an American Resort Community*. New Brunswick: Rutgers University Press, 1992.

Doughty, Robin W. "Sea Turtles in Texas: A Forgotten Commerce." *Southwestern Historical Quarterly* 88, no. 1 (1984): 43–70.

Dresser, H. E. "On the Late Dr. Walter's Ornithological Researches in the Taimyr Peninsula." *Ibis* 46, no. 2 (1904): 228–35.

Driggers, William B., III, G. Walter Ingram Jr., Mark A. Grace, Christopher T. Gledhill, Terry A. Henwood, Carrie N. Horton, and Christian M. Jones. "Pupping Areas and Mortality Rates of Young Tiger Sharks *Galeocerdo Cuvier* in the Western North Atlantic Ocean." *Aquatic Biology* 2, no. 2 (2008): 161–70.

Dubczak, J. "Proven Biomedical Horseshoe Crab Conservation Initiatives." Presentation at the International Workshop on the Science and Conservation of Asian Horseshoe Crabs, Hong Kong, June 12–16, 2011. In *Abstracts of Plenary Talks and Oral Presentations*, O–8. Hong Kong, 2011.

Duerr, Adam E., Bryan D. Watts, and Fletcher M. Smith. *Population Dynamics of Red Knots Stopping over in Virginia during Spring Migration*. Center for Conservation Biology Technical Report Series, CCBTR–11–04. Williamsburg: College of William and Mary and Virginia Commonwealth University, 2011.

Dunne, Pete. *Bayshore Summer: Finding Eden in a Most Unlikely Place*. Boston: Houghton Mifflin Harcourt, 2010.

———. "Knot Then, Knot Now, Knot Later." *Peregrine Observer* 34 (Summer 2012): 5–9.

———. *Tales of a Low-Rent Birder*. Austin: University of Texas Press, 1995.

Dunne, P., D. Sibley, C. Sutton, and W. Wander. "Aerial Surveys in Delaware Bay: Confirming an Enormous Spring Staging Area for Shorebirds." *Wader Study Group Bulletin* 35 (1982): 32–33.

Eagle, Josh. "Issues and Approaches in the Regulation of the Horseshoe Crab Fishery." In *Limulus in the Limelight*, 85–92. New York: Kluwer, 2001.

Eastern Massachusetts National Wildlife Refuge Complex. *Compatibility Determination*. Sudbury, Mass.: Eastern Massachusetts National Wildlife Refuge Complex, May 22, 2002.

Egevang, Carsten, Iain J. Stenhouse, Richard A. Phillips, Aevar Petersen, James W. Fox, and Janet R. D. Silk. "Tracking of Arctic Terns Sterna Paradisaea Reveals Longest Animal Migration." *Proceedings of the National Academy of Sciences*, February 2, 2010, 2078–81.

Ekblaw, W. Elmer. "Finding the Nest of the Knot." *Wilson Bulletin*, December 1, 1918, 97–100.

Elliott, Kyle Hamish, and Anthony J. Gaston. "Mass-Length Relationships and Energy Content of Fishes and Invertebrates Delivered to Nestling Thick-Billed Murres

Uria Lomvia in the Canadian Arctic, 1981–2007." *Marine Ornithology* 36 (n.d.): 25–33.

Emerson, Eva. "The Intrigue and Reach of Epigenetics Grows." *Science News* 183, no. 7 (2013): 2.

Escudero, Graciela, Juan G. Navedo, Theunis Piersma, Petra De Goeij, and Pim Edelaar. "Foraging Conditions 'at the End of the World' in the Context of Long-Distance Migration and Population Declines in Red Knots." *Austral Ecology*, May 1, 2012, 355–64.

Evenden, Matthew D. "The Laborers of Nature: Economic Ornithology and the Role of Birds as Agents of Biological Pest Control in North American Agriculture, ca. 1880–1930." *Forest and Conservation History*, October 1, 1995, 172–83.

Fanning, Mary M., Ron Wassel, and Toni Piazza-Hepp. "Pyrogenic Reactions to Gentamicin Therapy." *New England Journal of Medicine* 343, no. 22 (2000): 1658–59.

Feilden, H. W. "Breeding of the Knot in Grinnell Land." *British Birds* 13, no. 11 (1920): 278–82.

———. "List of Birds Observed in Smith Sound and in the Polar Basin during the Arctic Expedition of 1875–76." *Ibis* 19, no. 4 (1877): 401–12.

Feng, Song, Chang-Hoi Ho, Qi Hu, Robert J. Oglesby, Su-Jong Jeong, and Baek-Min Kim. "Evaluating Observed and Projected Future Climate Changes for the Arctic Using the Köppen-Trewartha Climate Classification." *Climate Dynamics*, April 1, 2012, 1359–73.

Ferrari, S., B. Ercolano, and C. Albrieu. "Pérdida de hábitat por actividades antrópicas en las marismas y planicies de marea del estuario del Río Gallegos (Patagonia Austral, Argentina)." In *Gestión sostenible de humedales: 3*, edited by M. Castro Lucic and Reyes Fernández, 19–327. Santiago: CYTED y Programa Internacional de Interculturalidad, 2007.

Ferrari, Silvia, Santiago Imberti, and Carlos Albrieu. "Magellanic Plovers *Pluvianellus Socialis* in Southern Santa Cruz Province, Argentina." *Wader Study Group Bulletin.* 101–2 (2003): 1–6.

Fisher, Robert A., and Dylan Lee Fisher. *The Use of Bait Bags to Reduce the Need for Horseshoe Crab as Bait in the Virginia Whelk Fishery.* VIMS Marine Resource Report No. 2006–10. Gloucester Point, Va.: Virginia Sea Grant, October 2006.

Fleckenstein, Henry, A., Jr. *Shorebird Decoys.* Exton, Pa.: Schiffer, 1980.

Flores, Marcelo M. A. *Antecedentes sobre la avifauna y mastozoofauna marina de Isla Riesco y áreas adyacentes.* Oceana, 2011.

Forbush, Edward Howe. *A History of the Game Birds, Wild-fowl and Shore Birds of Massachusetts and Adjacent States.* Boston: Massachusetts State Board of Agriculture, 1912.

Foster, Charles R., Anthony F. Amos, and Lee A. Fuiman. "Trends in Abundance of Coastal Birds and Human Activity on a Texas Barrier Island over Three Decades." *Estuaries and Coasts* 32, no. 6 (2009): 1079–89.

Fowler, Henry W. "The King Crab Fisheries in Delaware Bay." In *Annual Report of the New Jersey State Museum: Including a List of the Specimens Received during the Year, 1907*, 111–19. Trenton: MacCrellish & Quigley, 1908.

Fraser, J. D., S. M. Karpanty, J. B. Cohen, and B. R. Truitt. "The Red Knot (*Calidris Canutus Rufa*) Decline in the Western Hemisphere: Is There a Lemming Connection?" *Canadian Journal of Zoology* 91, no. 1 (2013): 13–16.

"Fresh Water Peril Seen in Rising Sea." *New York Times*, November 21, 1953, 15.

Friðriksson, Sturla, and Borgþór Magnússon. "Colonization of the Land." *Surtsey— The Surtsey Research Society*, 2007. http://www.surtsey.is/pp_ens/biola_1.htm.

Friedman, Richard L. "Aseptic Processing Contamination Case Studies and the Pharmaceutical Quality System." *PDA Journal of Pharmaceutical Science and Technology* 59 (April 2005): 118–26.

"From Cape May." *Philadelphia Inquirer*, August 10, 1853, 2.

Galetti, Mauro, and Rodolfo Dirzo. "Ecological and Evolutionary Consequences of Living in a Defaunated World." *Biological Conservation* 163 (2013): 1–6.

Gascoigne, William R., Dana Hoag, Lynne Koontz, Brian A. Tangen, Terry L. Shaffer, and Robert A. Gleason. "Valuing Ecosystem and Economic Services across Land-Use Scenarios in the Prairie Pothole Region of the Dakotas, USA." *Ecological Economics*, August 15, 2011, 1715–25.

Gaston, Anthony J., and Kyle H. Elliott. "Effects of Climate-Induced Changes in Parasitism, Predation and Predator-Predator Interactions on Reproduction and Survival of an Arctic Marine Bird." *Arctic*, August 3, 2013, 43–51.

Gaston, Anthony J., Paul A. Smith, and Jennifer F. Provencher. "Discontinuous Change in Ice Cover in Hudson Bay in the 1990s and Some Consequences for Marine Birds and Their Prey." *ICES Journal of Marine Science: Journal du conseil*, September 1, 2012, 1218–25.

Gauvry, G., and M. D. Janke. "Current Horseshoe Crab Harvesting Practices Cannot Support Global Demand for TAL/LAL." Presentation at the International Workshop on the Science and Conservation of Asian Horseshoe Crabs, Hong Kong, June 12–16, 2011. In *Abstracts of Plenary Talks and Oral Presentations*, PT-4. Hong Kong, 2011.

Gaylord, Brian, Tessa M. Hill, Eric Sanford, Elizabeth A. Lenz, Lisa A. Jacobs, Kirk N. Sato, Ann D. Russell, and Annaliese Hettinger. "Functional Impacts of Ocean Acidification in an Ecologically Critical Foundation Species." *Journal of Experimental Biology* 214, no. 15 (n.d.): 2586–94.

Germano, Frank. "Horseshoe Crabs: Balanced Management Plan Yields Fishery and Biomedical Benefits." *DMF News*, 2nd quarter (2003): 6–8.

Giaccardi, Maricel, and Laura M. Reyes, eds. *Plan de manejo del área natural protejida Bahía de San Antonio, Río Negro*. Gobierno de la Provincia de Río Negro, 2012.

Gibson, Rosemary, and Allan Baker. "Multiple Gene Sequences Resolve Phylogenetic Relationships in the Shorebird Suborder Scolopaci (Aves: Charadriiformes)." *Molecular Phylogenetics and Evolution* 64, no. 1 (2012): 66–72.

Gill, Robert E., Pablo Canevari, and Eve H. Iverson. "Eskimo Curlew (*Numenius Borealis*)." *Birds of North America Online*, 1998. http://bna.birds.cornell.edu/bna/species/347.

Gill, Robert E., Jr., David C. Douglas, Colleen M. Handel, T. Lee Tibbitts, Gary Hufford, and Theunis Piersma. "Hemispheric-Scale Wind Selection Facilitates Bar-Tailed Godwit Circum-Migration of the Pacific." *Animal Behaviour* 90 (2014): 117–30.

Given, Aaron M. "Leucistic Red Knot *Calidris Canutus* at Kiawah Island, South Carolina." *Wader Study Group Bulletin* 118, no. 1 (2011): 65.

Gobler, Christopher J., Elizabeth L. DePasquale, Andrew W. Griffith, and Hannes Baumann. "Hypoxia and Acidification Have Additive and Synergistic Negative Effects on the Growth, Survival, and Metamorphosis of Early Life Stage Bivalves." *PLoS ONE*, January 8, 2014, e83648.

González, Patricia M. "Las aves migratorias." In *Las mesetas patagónicas que caen al mar: La costa rionegrina*, edited by Ricardo Freddy Masera, Juana Lew, and Guillermo Serra Peirano, 321–48. Viedma: Gobierno de Río Negro, 2005.

González, Patricia M., Allan J. Baker, and María Eugenia Echave. "Annual Survival of Red Knots (*Calidris Canutus Rufa*) Using the San Antonio Oeste Stopover Site Is Reduced by Domino Effects Involving Late Arrival and Food Depletion in Delaware Bay." *Hornero* 21, no. 2 (2006): 109–17.

González, Patricia, Theunis Piersma, and Yvonne Verkuil. "Food, Feeding, and Refueling of Red Knots during Northward Migration at San Antonio Oeste, Rio Negro, Argentina." *Journal of Field Ornithology* 67, no. 4 (1996): 575–91.

Goode, George Brown. *The Fisheries and Fishery Industries of the United States.* Section 1, *Natural History of Useful Aquatic Animals.* Washington, D.C.: U.S. Commission of Fish and Fisheries, 1884.

"The Great King Crab Invasion." *Chicago Tribune*, July 24, 1871.

Greely, Adolphus Washington. *Three Years of Arctic Service: An Account of the Lady Franklin Bay Expedition of 1881–84.* Vol. 2. London: Richard Bentley & Son, 1886.

Green, Andy J., and Johan Elmberg. "Ecosystem Services Provided by Waterbirds." *Biological Reviews* (2013).

Green, Andy J., Jordi Figuerola, and Marta I. Sánchez. "Implications of Waterbird Ecology for the Dispersal of Aquatic Organisms." *Acta Oecologica* 23, no. 3 (2002): 177–89.

Greenberg, Joel. *A Feathered River across the Sky.* New York: Bloomsbury, 2014.

Hall, Ansley. "Notes on the Oyster Industry of New Jersey." In *Part VIII: Report of the Commissioner for the Year Ending June 30, 1892*, edited by U.S. Commission of Fish and Fisheries, 463–528. Washington, D.C.: U.S. Commission of Fish and Fisheries, 1894.

Hallmann, Caspar A., Ruud P. B. Foppen, Chris A. M. van Turnhout, Hans de Kroon, and Eelke Jongejans. "Declines in Insectivorous Birds Are Associated with High Neonicotinoid Concentrations." *Nature* 511, no. 7509 (2014): 341–43.

Hann, Roy W. "Fate of Oil from the Supertanker *Metula. International Oil Spill Conference Proceedings* 1 (March 1977): 465–68.

———. *VLCC* Metula *Oil Spill.* Washington, D.C.: U.S. Coast Guard, 1974.

Hannan, Laura B., James D. Roth, Llewellyn M. Ehrhart, and John F. Weishampel. "Dune Vegetation Fertilization by Nesting Sea Turtles." *Ecology,* April 1, 2007, 1053–58.

Hapgood, Warren, and Robert B. Roosevelt. *Shorebirds.* New York: Forest & Stream, 1881.

Haramis, G. Michael, A. Link, C. Osenton, David B. Carter, Richard G. Weber, Nigel A. Clark, Mark A. Teece, and David S. Mizrahi. "Stable Isotope and Pen Feeding Trial Studies Confirm the Value of Horseshoe Crab *Limulus Polyphemus* Eggs to Spring Migrant Shorebirds in Delaware Bay." *Journal of Avian Biology* 38, no. 3 (2007): 367–76.

Harrington, Brian. *The Flight of the Red Knot.* New York: W. W. Norton, 1996.

Harrington, B. A., and R. I. G. Morrison. "Notes on the Wintering Areas of Red Knot *Calidris Canutus Rufa* in Argentina, South America." *Wader Study Group Bulletin* 28 (1980): 40–42.

Harting, J. E. "Discovery of the Eggs of the Knot." *Zoologist* 9, no. 105 (1885): 344–45.

Henkel, Jessica R., Bryan J. Sigel, and Caz M. Taylor. "Large-Scale Impacts of the Deepwater Horizon Oil Spill: Can Local Disturbance Affect Distant Ecosystems through Migratory Shorebirds?" *BioScience* 62, no. 7 (2012): 676–85.

Hetland, Robert D., and Lisa Campbell. "Convergent Blooms of *Karenia Brevis* along the Texas Coast." *Geophysical Research Letters* 34, no. 19 (2007): L19604.

Hicklin, Peter W., and John W. Chardine. "The Morphometrics of Migrant Semipalmated Sandpipers in the Bay of Fundy: Evidence for Declines in the Eastern Breeding Population." *Waterbirds* 35, no. 1 (2012): 74–82.

Hildebrand, H. "Hallazgo del área de anidación de al tortouga 'Lora' *Lepidochelys Kempii* (Garman), en la costa occidental del Golfo de México (Reptila, Chelonia)." *Ciencia* (Mexico) 22 (1963): 105–112. Translated by Charles W. Caillouet Jr. and reproduced at http://www.seaturtle.org/mtn/archives.

Hladik, Michelle L., Dana W. Kolpin, and Kathryn M. Kuivila. "Widespread Occurrence of Neonicotinoid Insecticides in Streams in a High Corn and Soybean Producing Region, USA." *Environmental Pollution* 193 (October 2014): 189–96.

Hönisch, Bärbel, Andy Ridgwell, Daniela N. Schmidt, Ellen Thomas, Samantha J. Gibbs, Appy Sluijs, Richard Zeebe, et al. "The Geological Record of Ocean Acidification." *Science* 335, no. 6072 (2012): 1058–63.

Hope, C. E., and T. M. Shortt. "Southward Migration of Adult Shorebirds on the West Coast of James Bay, Ontario." *Auk* 61, no. 4 (1944): 572–76.

Hornaday, William Temple. *Our Vanishing Wildlife.* New York: New York Zoological Society, 1913.

Houghton, Jonathan D. R., Thomas K. Doyle, Mark W. Wilson, John Davenport, and Graeme C. Hays. "Jellyfish Aggregations and Leatherback Turtle Foraging

Patterns in a Temperate Coastal Environment." *Ecology*, August 1, 2006, 1967–72.

Howard-Jones, Norman. "Cholera Therapy in the Nineteenth Century." *Journal of the History of Medicine and Allied Sciences* 27, no. 4 (1972): 373–95.

Hu, M. H., Y. J. Wang, S. G. Cheung, P. K. S. Shin, and Q. Z. Li. "Distribution, Abundance and Population Structure of Horseshoe Crabs along Three Intertidal Zones of Beibu Gulf, Southern China." Presentation at the International Workshop on the Science and Conservation of Asian Horseshoe Crabs, Hong Kong, June 12–16, 2011. In *Abstracts of Plenary Talks and Oral Presentations*, O–1. Hong Kong, 2011.

Hunt, Harrison J., and Ruth Hunt Thompson. *North to the Horizon: Searching for Peary's Crocker Land*. Camden, Me.: Down East Books, 1980.

Hurton, Lenka, Jim Berkson, and Stephen Smith. "The Effect of Hemolymph Extraction Volume and Handling Stress on Horseshoe Crab Mortality." In *Biology and Conservation of Horseshoe Crabs*, edited by John T. Tanacredi, Mark L. Botton, and David R. Smith, 331–46. Dordrecht: Springer, 2009.

IBA Canada. "Churchill and Vicinity." *IBA Canada*. http://www.ibacanada.ca/site.jsp?siteID—B003&lang=EN.

IGBP, IOC, and SCOR. *Ocean Acidification Summary for Policymakers—Third Symposium on the Ocean in a High-CO2 World*. Stockholm: International Geosphere-Biosphere Programme, 2013.

Isack, H. A., and H.-U. Reyer. "Honeyguides and Honey Gatherers: Interspecific Communication in a Symbiotic Relationship." *Science* 243, no. 4896 (1989): 1343–46.

Iverson, Samuel, H. Grant Gilchrist, Paul Smith, Anthony J. Gaston, and Mark Forbes. "Longer Ice-Free Seasons Increase the Risk of Nest Depredation by Polar Bears for Colonial Breeding Birds in the Canadian Arctic." *Proceedings of the Royal Society B* 281, no. 1779 (2014).

Jack, Janis Graham. "Memorandum Opinion and Verdict of the Court: *The Aransas Project vs. Bryan Shaw*." U.S. District Court, Southern District of Texas, Corpus Christi Division, 2013.

Jackson, Jerome A. *George Miksch Sutton: Artist, Scientist, and Teacher*. Norman: University of Oklahoma Press, 2007.

James-Pirri, Mary-Jane. *Assessment of Spawning Horseshoe Crabs (Limulus Polyphemus) at Cape Cod National Seashore, 2008–2009*. Natural Resource Technical Report NPS/CACO/NRTR–2012/573. Fort Collins, Colo.: National Park Service, April 2012.

James-Pirri, M. J., K. Tuxbury, S. Marino, and S. Koch. "Spawning Densities, Egg Densities, Size Structure, and Movement Patterns of Spawning Horseshoe Crabs, *Limulus Polyphemus*, within Four Coastal Embayments on Cape Cod, Massachusetts." *Estuaries* 28, no. 2 (2005): 296–313.

Jansson, Anna, Joanna Norkko, and Alf Norkko. "Effects of Reduced pH on *Macoma Balthica* Larvae from a System with Naturally Fluctuating pH-Dynamics." *PloS One* 8, no. 6 (2013): e68198.

Jehl, Joseph R., Jr. "Disappearance of Breeding Semipalmated Sandpipers from Churchill, Manitoba: More than a Local Phenomenon." *Condor* 109, no. 2 (2007): 351–60.

———. "*Pluvianellus Socialis:* Biology, Ecology and Relationships of an Enigmatic Patagonian Shorebird." *Transactions of the San Diego Society of Natural History* 18 (1975): 25–74.

Jenkins, Scott, Jeffrey Paduan, Philip Roberts, Daniel Schlenk, and Judith Weiss. *Management of Brine Discharges to Coastal Waters: Recommendations of a Science Advisory Panel.* Costa Mesa: Southern California Coastal Water Research Project, 2013.

Jetz, W., G. H. Thomas, J. B. Joy, K. Hartmann, and A. O. Mooers. "The Global Diversity of Birds in Space and Time." *Nature*, November 15, 2012, 444–48.

Johnson, A. W. *The Birds of Chile and Adjacent Regions of Argentina, Bolivia, and Peru.* Vol. 1. Buenos Aires: Platt Establecimientos Gráficos, 1965.

Johnson, Michael A., Paul I. Padding, Michel H. Gendron, Eric T. Reed, and David A. Graber. "Assessment of Harvest from Conservation Actions for Reducing Midcontinent Light Geese and Recommendations for Future Monitoring." In *Evaluation of Special Management Measures for Midcontinent Lesser Snow Geese and Ross's Geese*, edited by J. O. Leafloor, T. J. Moser, and B. D. J. Batt, 46–94. Arctic Goose Joint Venture Special Publication. Washington, D.C., and Ottawa: U.S. Fish and Wildlife Service and Canadian Wildlife Service, 2012.

Jones, Sierra J., Fernando P. Lima, and David S. Wethey. "Rising Environmental Temperatures and Biogeography: Poleward Range Contraction of the Blue Mussel, *Mytilus Edulis* L., in the Western Atlantic." *Journal of Biogeography* 37, no. 12 (2010): 2243–59.

Kanchanapongkul, Jirasak. "Tetrodotoxin Poisoning Following Ingestion of the Toxic Eggs of the Horseshoe Crab *Carcinoscorpius Rotundicauda*: A Case Series from 1994 through 2006." *Southeast Asian Journal of Tropical Medicine and Public Health* 39, no. 2 (2008): 303–6.

Kanchanapongkul, J., and P. Krittayapoositpot. "An Epidemic of Tetrodotoxin Poisoning Following Ingestion of the Horseshoe Crab *Carcinoscorpius Rotundicauda*." *Southeast Asian Journal of Tropical Medicine and Public Health* 26, no. 2 (1995): 364–67.

Karp, Daniel S., Chase D. Mendenhall, Randi Figueroa Sandí, Nicolas Chaumont, Paul R. Ehrlich, Elizabeth A. Hadly, and Gretchen C. Daily. "Forest Bolsters Bird Abundance, Pest Control and Coffee Yield." *Ecology Letters* 16, no. 11 (2013): 1339–47.

Kaufman, Kenn. *Lives of North American Birds.* Boston: Houghton Mifflin, 1996.

Kays, Roland, Patrick A. Jansen, Elise M. H. Knecht, Reinhard Vohwinkel, and Martin Wikelski. "The Effect of Feeding Time on Dispersal of Virola Seeds by Toucans Determined from GPS Tracking and Accelerometers." *Acta Oecologica* 37, no. 6 (2011): 625–31.

Kerbes, R. H., K. M. Meeres, and R. T. Alisaukas. *Surveys of Nesting Lesser Snow Geese and Ross's Geese in Arctic Canada, 2002–2009.* Arctic Goose Joint Venture

Special Publication. Washington, D.C., and Ottawa: U.S. Fish and Wildlife Service and Canadian Wildlife Service, 2014.

Kilpatrick, A. Marm. "Globalization, Land Use, and the Invasion of West Nile Virus." *Science*, October 21, 2011, 323–27.

Kimmelman, Michael. "Former Landfill, a Park to Be, Proves a Savior in the Hurricane." *New York Times*, December 17, 2012, C5.

King, F. Wayne. "Historical Review of the Decline of the Green Turtle and Hawksbill." In *Biology and Conservation of Sea Turtles: Proceedings of the World Conference on Sea Turtle Conservation. Washington, D.C., 26–30 November, 1979*, edited by Karen Bjorndal, 183–88. Washington, D.C.: Smithsonian Institution Press, 1995.

King, P. P. *Voyages of the* Adventure *and* Beagle. Vol. 1. London: Henry Colburn, 1839.

Krauss, Scott, Caroline A. Obert, John Franks, David Walker, Kelly Jones, Patrick Seiler, Larry Niles, S. Paul Pryor, John C. Obenauer, and Clayton W. Naeve. "Influenza in Migratory Birds and Evidence of Limited Intercontinental Virus Exchange." *PLoS Pathogens* 3, no. 11 (2007): e167.

Krauss, Scott, David E. Stallknecht, Nicholas J. Negovetich, Lawrence J. Niles, Richard J. Webby, and Robert G. Webster. "Coincident Ruddy Turnstone Migration and Horseshoe Crab Spawning Creates an Ecological 'Hot Spot' for Influenza Viruses." *Proceedings of the Royal Society B: Biological Sciences*, November 22, 2010, 3373–79.

Kurz, W., and M. J. James-Pirri. "The Impact of Biomedical Bleeding on Horseshoe Crab, *Limulus Polyphemus*, Movement Patterns on Cape Cod, Massachusetts." *Marine and Freshwater Behaviour and Physiology* 35, no. 4 (2002): 261–68.

LaDeau, Shannon L., A. Marm Kilpatrick, and Peter P. Marra. "West Nile Virus Emergence and Large-Scale Declines of North American Bird Populations." *Nature*, June 7, 2007, 710–13.

Lane, J. Perry. "Eels and Their Utilization." *Marine Fisheries Review*, April 1978, 1–20.

Latorre, Claudio, Patricio I. Moreno, Gabriel Vargas, Antonio Maldonado, Rodrigo Villa-Martínez, Juan J. Amresto, Carolina Villagrán, Mario Pino, Lauraro Núñez, and Martin Grosjean. "Late Quaternary Environments and Palaeoclimate." In *The Geology of Chile*, edited by Teresa Moreno and Wes Gibbons, 309–28. London: Geological Society of London, 2007.

Leibovitz, Louis, and Gregory Lewbart. "Diseases and Symbionts: Vulnerability Despite Tough Shells." In *The American Horseshoe Crab*, edited by Carl N. Shuster Jr., Robert B. Barlow, and H. Jane Brockmann, 245–75. Cambridge, Mass.: Harvard University Press, 2003.

Lelli, Barbara, David E. Harris, and AbouEl-Makarim Aboueissa. "Seal Bounties in Maine and Massachusetts, 1888 to 1962." *Northeastern Naturalist*, July 1, 2009, 239–54.

Lenes, J. M., B. A. Darrow, J. J. Walsh, J. M. Prospero, R. He, R. H. Weisberg, G. A. Vargo, and C. A. Heil. "Saharan Dust and Phosphatic Fidelity: A Three-Dimensional Biogeochemical Model of *Trichodesmium* as a Nutrient Source for Red Tides on the West Florida Shelf." *Continental Shelf Research* 28, no. 9 (2008): 1091–1115.

Leopold, Aldo. *A Sand County Almanac: With Other Essays on Conservation from Round River*. New York: Random House, 1966.

Leschen, A. S., and S. J. Correia. "Mortality in Female Horseshoe Crabs (*Limulus Polyphemus*) from Biomedical Bleeding and Handling: Implications for Fisheries Management." *Marine and Freshwater Behaviour and Physiology* 43, no. 2 (2010): 135–47.

————. "Response to Associates of Cape Cod Comments on 'Mortality in Female Horseshoe Crabs (*Limulus Polyphemus*) from Biomedical Bleeding and Handling: Implications for Fisheries Management' by A. S. Leschen and S. J. Correia (2010)," November 2, 2010.

Levere, Trevor Harvey. *Science and the Canadian Arctic: A Century of Exploration, 1818–1918*. Cambridge: Cambridge University Press, 1993.

Levin, Jack. "The History of the Development of the Limulus Amebocyte Lysate Test." In *Bacterial Endotoxins: Structure, Biomedical Significance, and Detection with the Limulus Amebocyte Lysate Test*, edited by Harry R. Büller, Augueste Sturk, Jack Levin, and Jan W. ten Cate, 3–28. New York: Alan R. Liss, 1985.

Levin, Jack, H. Donald Hochstein, and Thomas J. Novitsky. "Clotting Cells and Limulus Amebocyte Lysate: An Amazing Analytical Tool." In *The American Horseshoe Crab*, edited by Carl N. Shuster Jr., Robert B. Barlow, and H. Jane Brockmann, 310–40. Cambridge, Mass.: Harvard University Press, 2003.

Leyrer, Jutta, Tamar Lok, Maarten Brugge, Anne Dekinga, Bernard Spaans, Jan A. van Gils, Brett K. Sandercock, and Theunis Piersma. "Small-Scale Demographic Structure Suggests Preemptive Behavior in a Flocking Shorebird." *Behavioral Ecology* 23, no. 6 (2012): 1226–33.

Lomax, Dean R., and Christopher A. Racay. "A Long Mortichnial Trackway of *Mesolimulus walchi* from the Upper Jurassic Solnhofen Lithographic Limestone near Wintershof, Germany." *Ichnos* 19, no. 3 (2012): 175–83.

Loveland, Robert E. "The Life History of Horseshoe Crabs." In *Limulus in the Limelight*, edited by John T. Tanacredi, 93–101. New York: Kluwer, 2001.

Loverock, Bruce, Barry Simon, Allen Burgenson, and Alan Baines. "A Recombinant Factor C Procedure for the Detection of Gram-Negative Bacterial Endotoxina." *Pharmacopeial Forum* 36, no. 1 (2010): 321–29.

Lynam, Christopher P., Mark J. Gibbons, Bjørn E. Axelsen, Conrad A.J. Sparks, Janet Coetzee, Benjamin G. Heywood, and Andrew S. Brierley. "Jellyfish Overtake Fish in a Heavily Fished Ecosystem." *Current Biology* 16, no. 13 (2006): R492–93.

Machut, L. S., and K. E. Limburg. "*Anguillicola Crassus* Infection in *Anguilla Rostrata* from Small Tributaries of the Hudson River Watershed, New York, USA." *Diseases of Aquatic Organisms* 79, no. 1 (2007): 37–45.

Mackay, George H. "Observations on the Knot (*Tringa Canutus*)." *Auk* 10, no. 1 (1893): 25–35.

MacKenzie, Clyde L. "History of the Fisheries of Raritan Bay, New York and . . ." *Marine Fisheries Review* 52, no. 4 (1990): 1–45.

MacKinnon, J. B. *The Once and Future World*. Boston: Houghton Mifflin, 2013.

MacMillan, Donald Baxter. *How Peary Reached the Pole: The Personal Story of His Assistant, Donald B. MacMillan* . . . Boston: Houghton Mifflin, 1934.

Magaña, Hugo A., Cindy Contreras, and Tracy A. Villareal. "A Historical Assessment of *Karenia Brevis* in the Western Gulf of Mexico." *Harmful Algae* 2, no. 3 (2003): 163–71.

Magnússon, Borgthór, Sigurdur H. Magnússon, and Sturla Friðriksson. "Developments in Plant Colonization and Succession on Surtsey during 1999–2008." *Surtsey Research* 12 (2009): 57–76.

Mallory, M. L., A. J. Gaston, H. G. Gilchrist, G. J. Robertson, and B. M. Braune. "Effects of Climate Change, Altered Sea-Ice Distribution and Seasonal Phenology on Marine Birds." In *A Little Less Arctic*, edited by Steven H. Ferguson, Lisa L. Loseto, and Mark L. Mallory, 179–95. Dordrecht: Springer Netherlands, 2010.

Manikkam, Mohan, Rebecca Tracey, Carlos Bosagna-Guerrero, and Michael K. Skinner. "Plastics Derived Endocrine Disruptors (BPA, DEHP and DBP) Induce Epigenetic Transgenerational Inheritance of Obesity, Reproductive Disease and Sperm Epimutations." *PLoS ONE*, January 24, 2013.

Mann, Michael E. *The Hockey Stick and the Climate Wars*. New York: Columbia University Press, 2012.

Manning, T. H. "Some Notes on Southampton Island." *Geographical Journal* 88, no. 3 (1936): 232–242.

Manville, Alfred M. "Framing the Issues Dealing with Migratory Birds, Commercial Land-Based Wind Energy Development, USFWS, and the MBTA." Presentation at the conference on the Migratory Bird Treaty Act Lewis and Clark Law School, October 21, 2011.

Markandya, Anil, Tim Taylor, Alberto Longo, M. N. Murty, Sucheta Murty, and K. Dhavala. "Counting the Cost of Vulture Decline—An Appraisal of the Human Health and Other Benefits of Vultures in India." *Ecological Economics* 67, no. 2 (2008): 194–204.

Marsh, Christopher P., and Philip M. Wilkinson. "Significance of the Central Coast of South Carolina as Critical Shorebird Habitat." *Chat* 54 (Fall 1991): 69–92.

Martini, I. P., and R. I. G. Morrison. "Regional Distribution of *Macoma balthica* and *Hydrobia minuta* on the Subarctic Coasts of Hudson Bay and James Bay, Ontario, Canada." *Estuarine, Coastal and Shelf Science* 24, no. 1 (1987): 47–68.

Martinic, Mateo. *Brief History of the Land of Magellan*. Translated by Juan C. Judikis. Punta Arenas: Universidad de Magallanes, 2002.

Mathew, W. M. "Peru and the British Guano Market, 1840–1870." *Economic History Review* 23, no. 1 (1970): 112.

Matthiessen, Peter. "Happy Days." *Audubon*, November 1975, 64–95.

Maxted, Angela M., M. Page Luttrell, Virginia H. Goekjian, Justin D. Brown, Lawrence J. Niles, Amanda D. Dey, Kevin S. Kalasz, David E. Swayne, and David E. Stallknecht. "Avian Influenza Virus Infection Dynamics in Shorebird Hosts." *Journal of Wildlife Diseases* 48 (2012): 322–34.

Maxted, Angela M., Ronald R. Porter, M. Page Luttrell, Virginia H. Goekjian, Amanda D. Dey, Kevin S. Kalasz, Lawrence J. Niles, and David E. Stallknecht. "Annual Survival of Ruddy Turnstones Is Not Affected by Natural Infection with Low Pathogenicity Avian Influenza Viruses." *Avian Diseases* 56, no. 3 (2012): 567–73.

McCauley, Douglas J. "Selling Out on Nature." *Nature* 443 (2006): 27–28.

McCauley, Douglas J., Paul A. DeSalles, Hillary S. Young, Robert B. Dunbar, Rodolfo Dirzo, Matthew M. Mills, and Fiorenza Micheli. "From Wing to Wing: The Persistence of Long Ecological Interaction Chains in Less-Disturbed Ecosystems." *Nature Scientific Reports* 2, (2012): 1–5.

McClenachan, Loren, Jeremy B. C. Jackson, and Marah J. H. Newman. "Conservation Implications of Historic Sea Turtle Nesting Beach Loss." *Frontiers in Ecology and the Environment* 4, no. 6 (2006): 290–96.

McDonald, Colin. "Wind Farms and Deadly Skies." *San Antonio Express News,* February 26, 2011. http://www.mysanantonio.com/living_green_sa/article/Wind-farmsand-deadly-skies–1032765.php.

McDonald, Marshall. "Fisheries of the Delaware River." In *The Fisheries and Fisheries Industries of the United States,* section 5, vol. 1, *Histories and Methods of the Fisheries,* edited by George Brown Goode, 654–57. Washington, D.C.: U.S. Commission of Fish and Fisheries, 1887.

McKinnon, L., P. A. Smith, E. Nol, J. L. Martin, F. I. Doyle, K. F. Abraham, H. G. Gilchrist, R. I. G. Morrison, and J. Bêty. "Lower Predation Risk for Migratory Birds at High Latitudes." *Science* 327, no. 5963 (2010): 326–27.

McPhee, John. *The Founding Fish.* New York: Farrar, Straus & Giroux, 2002.

Merriam, C. Hart. "The Eggs of the Knot (*Tringa Canutus*) Found at Last!" *Auk,* July 1, 1885, 312–13.

Meyer de Schauensee, Rodolphe. *The Species of Birds of South America and Their Distribution.* Philadelphia: Academy of Natural Sciences, 1966.

Michaels, David. *Doubt Is Their Product.* New York: Oxford University Press, 2008.

Migratory Bird Conservation Commission. *2012 Annual Report.* U.S. Fish and Wildlife Service, 2013.

Milius, Susan. "Cat-Induced Death Toll Revised." *Science News,* March 8, 2014, 30.

———. "Windows Are Major Bird Killers." *Science News,* March 22, 2014, 8–9.

Millam, Doris. "The History of Intravenous Therapy." *Journal of Infusion Nursing* 19, no. 1 (1996): 5–15.

Miller, Gifford H., Scott J. Lehman, Kurt A. Refsnider, John R. Southon, and Yafang Zhong. "Unprecedented Recent Summer Warmth in Arctic Canada." *Geophysical Research Letters* 40, no. 21 (2013): 5745–51.

Miller, Kenneth G., Robert E. Kopp, Benjamin P. Horton, James V. Browning, and Andrew C. Kemp. "A Geological Perspective on Sea-Level Rise and Its Impacts along the U.S. Mid-Atlantic Coast." *Earth's Future,* December 1, 2013, 3–18.

Mineau, Pierre, and Cynthia Palmer. *The Impact of the Nation's Most Widely Used Insecticides on Birds.* American Bird Conservancy, 2013.

Mineau, Pierre, and Melanie Whiteside. "Pesticide Acute Toxicity Is a Better Correlate of US Grassland Bird Declines than Agricultural Intensification." *PloS One* 8, no. 2 (2013): e57457.

Mizrahi, David S., and Kimberly A. Peters. "Relationships between Sandpipers and Horseshoe Crab in Delaware Bay: A Synthesis." In *Biology and Conservation of Horseshoe Crabs*, edited by John T. Tanacredi, Mark L. Botton, and David R. Smith, 65–87. New York: Springer, 2009.

Mizrahi, David S., Kimberly A. Peters, and Patricia A. Hodgetts. "Energetic Condition of Semipalmated and Least Sandpipers during Northbound Migration Staging Periods in Delaware Bay." *Waterbirds* 35, no. 1 (2012): 135–45.

Moczek, Armin P., Sonia Sultan, Susan Foster, Cris Ledón-Rettig, Ian Dworkin, H. Fred Nijhout, Ehab Abouheif, and David W. Pfennig. "The Role of Developmental Plasticity in Evolutionary Innovation." *Proceedings: Biological Sciences / The Royal Society* 278, no. 1719 (2011): 2705–13.

Molnár, Péter K., Andrew E. Derocher, Tin Klanjscek, and Mark A. Lewis. "Predicting Climate Change Impacts on Polar Bear Litter Size." *Nature Communications*, February 8, 2011, 186.

Molnár, Péter K., Andrew E. Derocher, Gregory W. Thiemann, and Mark A. Lewis. "Predicting Survival, Reproduction and Abundance of Polar Bears under Climate Change." *Biological Conservation* 143, no. 7 (2010): 1612–22.

Monomoy National Wildlife Refuge. *Monomoy National Wildlife Refuge Draft Comprehensive Conservation Plan and Environmental Impact Statement.* Vols. 1–2. Sudbury, Mass.: U.S. Fish and Wildlife Service, April 2014.

Morris, Michael A. *The Strait of Magellan.* Dordrecht: Martinu Nijhoff, 1988.

Morrison, R. I. Guy, Nick C. Davidson, and Theunis Piersma. "Transformations at High Latitudes: Why Do Red Knots Bring Body Stores to the Breeding Grounds?" *Condor* 107, no. 2 (2005): 449–57.

Morrison, R. I. Guy, and Brain A. Harrington. "Critical Shorebird Resources in James Bay and Eastern North America." In *Transactions*, 498–506. Washington, D.C.: Wildlife Management Institute, 1979.

———. "The Migration System of the Red Knot *Calidris Canutus Rufa* in the New World." *Wader Study Group Bulletin* 64 (1992): 71–84.

Morrison, R.I. Guy, David S. Mizrahi, R. Kenyon Ross, Otte H. Ottema, Nyls de Pracontal, and Andy Narine. "Dramatic Declines of Semipalmated Sandpipers on Their Major Wintering Areas in the Guianas, Northern South America." *Waterbirds* 35, no. 1 (2012): 120–34.

Morrison, R. I. G., and R. K. Ross. *Atlas of Nearctic Shorebirds on the Coast of South America.* Vol. 1. Ottawa: Canadian Wildlife Service, 1989.

———. *Atlas of Nearctic Shorebirds on the Coast of South America.* Vol. 2. Ottawa: Canadian Wildlife Service, 1989.

Morrison, R. I. G., and Arie L. Spaans. "National Geographic Mini-Expedition to Surinam, 1978." *Wader Study Group Bulletin* 26 (1979): 37–41.

Morrison, Samuel Eliot. *The European Discovery of America: The Southern Voyages, A.D. 1492–1616.* New York: Oxford University Press, 1974.

Moser, Mary L., Wesley S. Patrick, John U. Crutchfield Jr., and W. L. Montgomery. "Infection of American Eels, *Anguilla Rostrata*, by an Introduced Nematode Parasite, *Anguillicola Crassus*, in North Carolina." *Copeia* 3 (2001): 848–53.

Mowat, Farley. *Sea of Slaughter*. Mechanicsburg, Pa.: Stackpole, 2004.

Murray, Molly. "Delaware Gets Millions to Help Beaches, Wetlands." Delawareonline. com, June 16, 2014.

Muston, Samuel. "Cafe de Mort: My Night Eating Dangerously." *Independent*, March 8, 2013. http://www.independent.ie/lifestyle/food-drink/cafe-de-mort-my-night-eating-dangerously–29117619.html.

Myers, J. P. "Sex and Gluttony on Delaware Bay." *Natural History* 95, no. 5 (1986): 68–77.

Myers, J. P., R. I. G. Morrison, Paolo Z. Antas, Brain A. Harrington, Thomas E. Lovejoy, Michel Sallaberry, Stanley E. Senner, and Arturo Tarak. "Conservation Strategy for Migratory Species." *American Scientist*, January 1, 1987, 18–26.

National Marine Fisheries Service (NMFS). *Atlantic Sturgeon New York Bight Distinct Population Segment: Endangered*. NMFS. http://www.nmfs.noaa.gov/pr/pdfs/species/atlanticsturgeon_nybright_dps.pdf.

National Oceanographic and Atmospheric Administration (NOAA). "Atlantic Coastal Fisheries Cooperative Management Act Provisions; Horseshoe Crab Fishery; Closed Area." *Federal Register*, February 5, 2001, 8906–11.

Nätt, Daniel, Niclas Lindqvist, Henrik Stranneheim, Joakim Lundeberg, Peter A. Torjesen, and Per Jensen. "Inheritance of Acquired Behaviour Adaptations and Brain Gene Expression in Chickens." *PLoS ONE*, July 28, 2009, e6405.

Neck, Raymond W. "Occurrence of Marine Turtles in the Lower Rio Grande of South Texas (Reptilia, Testudines)." *Journal of Herpetology* 12, no. 3 (1978): 422–27.

New Jersey Audubon, American Littoral Society, Delaware Riverkeeper Network, and the Conserve Wildlife Foundation of New Jersey. "Public Comments to the U.S. Fish and Wildlife Service." Docket ID FWS-R5-ES–2013–0097–0697, June 16, 2014.

New Jersey Geological Survey. *Geology of the County of Cape May, State of New Jersey*. Trenton: Printed at the Office of the True American, 1857.

Newman, Edward, ed. *A Dictionary of British Birds: Being a Reprint of Montagu's Ornithological Dictionary, Together with the Additional Species Described by Selby; Yarrell, in All Three Editions; and in Natural-History Journals*. London: W. Swan Sonnenschein & Allen, 1881.

Newstead, David J., Lawrence J. Niles, Ronald R. Porter, Amanda D. Dey, Joanna Burger, and Owen N. Fitzsimmons. "Geolocation Reveals Mid-continent Migratory Routes and Texas Wintering Areas of Red Knots *Calidris Canutus Rufa*." *Wader Study Group Bulletin* 120, no. 1 (2013): 53–59.

Newton, Alfred. "Abstract of Mr. J. Wolley's Researches in Iceland Respecting the Gare-Fowl or Great Auk (*Alea Impennis*, Linn.)." *Ibis* 3, no. 4 (1861): 374–99.

Ng, Sheau-Fang, Ruby C. Y. Lin, D. Ross Laybutt, Romain Barres, Julie A. Owens, and Margaret J Morris. "Chronic High-Fat Diet in Fathers Programs B-Cell Dysfunction in Female Rat Offspring." *Nature* 467, no. 7318 (2010): 963–66.

Ngy, Laymithuna, Chun-Fai Yu, Tomohiro Takatani, and Osamu Arakawa. "Toxicity Assessment for the Horseshoe Crab *Carcinoscorpius Rotundicauda* Collected from Cambodia." *Toxicon* 49, no. 6 (2007): 843–47.

Niles, Lawrence J. "What We Still Don't Know." *Rube with a View*, May 27, 2011. http://arubewithaview.com/2011/05/.

Niles, Lawrence, Joanna Burger, Ronald Porter, Amanda Dey, Stephanie Koch, Brian Harrington, Kate Iaquinto, and Matthew Boarman. "Migration Pathways, Migration Speeds and Non-breeding Areas Used by Northern Hemisphere Wintering Red Knots (*Calidris Canutus*) of the Subspecies Rufa." *Wader Study Group Bulletin* 119, no. 3 (2013): 195–203.

Niles, Lawrence J., Joanna Burger, Ronald R. Porter, Amanda D. Dey, Clive D. T. Minton, Patricia M. González, Allan J. Baker, James W. Fox, and Caleb Gordon. "First Results Using Light Level Geolocators to Track Red Knots in the Western Hemisphere Show Rapid and Long Intercontinental Flights and New Details of Migration Pathways." *Wader Study Group Bulletin* 117, no. 2 (2010): 123–30.

Niles, L. J., A. M. Smith, D. F. Daly, T. Dillingham, W. Shadel, A. D. Dey, M. S. Danihel, S. Hafner, and D. Wheeler. *Restoration of Horseshoe Crab and Migratory Shorebird Habitat on Five Delaware Bay Beaches Damaged by Superstorm Sandy.* Report to New Jersey Natural Lands Trust, December 27, 2013.

Nilsson, Eric, Ginger Larsen, Mohan Manikkam, Carlos Bosagna-Guerrero, Marina I. Savenkova, and Michael K. Skinner. "Environmentally Induced Epigenetic Transgenerational Inheritance of Ovarian Disease." *PLoS One*, May 3, 2012.

Nogales, Manuel, Félix M. Medina, Vicente Quilis, and Mercedes González-Rodríguez. "Ecological and Biogeographical Implications of Yellow-Legged Gulls (*Larus Cachinnans Pallas*) as Seed Dispersers of *Rubia Fruticosa* Ait. (Rubiaceae) in the Canary Islands." *Journal of Biogeography* 28, no. 9 (2001): 1137–45.

Nolet, Bart A., Silke Bauer, Nicole Feige, Yakov I. Kokorev, Igor Yu Popov, and Barwolt S. Ebbinge. "Faltering Lemming Cycles Reduce Productivity and Population Size of a Migratory Arctic Goose Species." *Journal of Animal Ecology* 82, no. 4 (2013): 804–13.

North American Bird Conservation Initiative Canada. *The State of Canada's Birds, 2012.* Ottawa: Environment Canada, 2012.

Novitsky, Thomas J. "Biomedical Applications of Limulus Amebocyte Lysate." In *Biology and Conservation of Horseshoe Crabs*, edited by John T. Tanacredi, Mark L. Botton, and David R. Smith, 315–29. Dordrecht: Springer, 2009.

Obmascik, Mark. *The Big Year: A Tale of Man, Nature, and Fowl Obsession.* New York: Free Press, 2004.

O'Brien, Michael, Richard Crossley, and Kevin Karlson. *The Shorebird Guide.* Boston: Houghton Mifflin, 2006.

O'Donnell, Michael J., Matthew N. George, and Emily Carrington. "Mussel Byssus Attachment Weakened by Ocean Acidification." *Nature Climate Change* 3, no. 6 (2013): 587–90.

Olinger, John Peter. "The Guano Age in Peru." *History Today* 30, no. 6 (1980): 13–18.

Oreskes, Naomi, and Erik M. Conway. *Merchants of Doubt*. New York: Bloomsbury, 2011.

Ostfeld, Richard S., Charles D. Canham, Kelly Oggenfuss, Raymond J. Winchcombe, and Felicia Keesing. "Climate, Deer, Rodents, and Acorns as Determinants of Variation in Lyme-Disease Risk." *PLoS Biology* 4, no. 6 (2006): e145.

Ottema, Otte H., and Arie L. Spaans. "Challenges and Advances in Shorebird Conservation in the Guianas, with a Focus on Suriname." *Ornitologia Neotropical*, supplement, 19 (2008): 339–46.

Owens, E. H. "Time Series Observations of Marsh Recovery and Pavement Persistence at Three Metula Spill Sites after 30 1/2 Years." *Proceedings of the 28th Arctic and Marine Oil Spill Programme (AMOP) Technical Seminar, Environment Canada* (2005): 463–72.

Pain, Deborah J., A. A. Cunningham, P. F. Donald, J. W. Duckworth, D. C. Houston, T. Katzner, J. Parry-Jones, C. Poole, V. Prakash, and P. Round. "Causes and Effects of Temporospatial Declines of Gyps Vultures in Asia." *Conservation Biology* 17, no. 3 (2003): 661–71.

"Palaeontology: Early Bird Was Black." *Nature*, February 9, 2012, 135.

"Parasitic Jaeger, Polar Bird of Prey, Seen Near Cape May." *New York Times*, July 25, 1922, 13.

Parmelee, David Freeland, H. A. Stephens, and Richard H. Schmidt. *The Birds of Southeastern Victoria Island and Adjacent Small Islands*. Bulletin 222. Ottawa: National Museum of Canada, 1967.

Parry, William Edward. *Appendix to Captain Parry's Journal of a Second Voyage for the Discovery of a North-west Passage from the Atlantic to the Pacific*. London: J. Murray, 1825. http://www.biodiversitylibrary.org/bibliography/48565.

———. *Journal of a Second Voyage for the Discovery of a North-west Passage from the Atlantic to the Pacific: Performed in the Years 1821–22–23*. London: J. Murray, 1824. http://archive.org/details/cihm_42230.

———. *Supplement to the Appendix of Captain Parry's Voyage for the Discovery of a North-west Passage in the Years 1819–20: Containing an Account of the Subjects of Natural History*. London: J. Murray, 1824. http://archive.org/details/cihm_39499.

Peacock, E., A. E. Derocher, N. J. Lunn, and M. E. Obbard. "Polar Bear Ecology and Management in Hudson Bay in the Face of Climate Change." In *A Little Less Arctic*, edited by Steven H. Ferguson, Lisa L. Loseto, and Mark L. Mallory, 93–116. Dordrecht: Springer Netherlands, 2010.

Perkins, Deborah E., Paul A. Smith, and H. Grant Gilchrist. "The Breeding Ecology of Ruddy Turnstones (*Arenaria Interpres*) in the Eastern Canadian Arctic." *Polar Record* 43, no. 02 (2007): 135–42.

Perovich, D., S. Gerland, S. Hendricks, W. Meier, M. Nicolaus, J. Richter-Menge, and M. Tschudi. "Sea Ice." *Arctic Report Card: Update for 2013*, November 21, 2013. http://www.arctic.noaa.gov/reportcard/exec_summary.html.

Pettingill, Olin Sewall, Jr. "In Memoriam: George Miskch Sutton." *Auk* 101 (January 1984): 146–52.

Pettis, Jeffery S., Elinor M. Lichtenberg, Michael Andree, Jennie Stitzinger, Robyn Rose, Dennis vanEngelsdorp, et al. "Crop Pollination Exposes Honey Bees to Pesticides Which Alters Their Susceptibility to the Gut Pathogen *Nosema Ceranae*." *PLoS ONE* 8, no. 7 (2013): e70182.

Phillips, Edward. *The New World of Words; or, Universal English Dictionary, Containing an Account of the Original or Proper Sense and Various Significations of All Hard Words Derived from Other Languages*. London: J. Phillips, 1720.

Piersma, Theunis. "Flyway Evolution Is Too Fast to Be Explained by the Modern Synthesis: Proposals for an 'Extended' Evolutionary Research Agenda." *Journal of Ornithology* 152, no. 1 (2011): 151–59.

Piersma, Theunis, and Jan A. van Gils. *The Flexible Phenotype*. Oxford: Oxford University Press, 2011.

Pilkey, Orrin, and Rob Young. *The Rising Sea*. Washington, D.C.: Island, 2009.

Pimentel, David. "Environmental and Economic Costs of the Application of Pesticides Primarily in the United States." In *Integrated Pest Management: Innovation-Development Process*, edited by Rajinder Peshin and Ashok K. Dhawan, 1:89–111. New York: Springer Science + Business Media, 2009.

Pimm, Stuart L. *The World According to Pimm: A Scientist Audits the Earth*. New York: McGraw-Hill, 2001.

Pirie, Lisa, Victoria Johnston, and Paul A. Smith. "Tier 2 Surveys." In *Arctic Shorebirds in North America: A Decade of Monitoring*, edited by J. Bart and V. Johnston, 185–94. Studies in Avian Biology 44. Berkeley: University of California Press, 2012.

Pleske, F. D. *Birds of the Eurasian Tundra*. Memoirs of the Boston Society of Natural History, vol. 6, no. 3. Boston: Boston Society of Natural History, 1928.

Pollock, Lisa A., Kenneth F. Abraham, and Erica Nol. "Migrant Shorebird Use of Akimiski Island, Nunavut: A Sub-Arctic Staging Site." *Polar Biology* 35 (2012): 1691–1701.

Potter, Julian K. "The Season." *Bird-lore* 36, no. 4 (1934): 242.

Powell, Cindie. "Water, Water, Everywhere." *Texas Shores* 40, no. 2 (2012): 11–29.

President's Task Force on Wildlife Diversity Funding. *Final Report*. Washington, D.C.: Association of Fish and Wildlife Agencies, September 1, 2011.

Prothero, Donald R. *Evolution: What the Fossils Say and Why It Matters*. New York: Columbia University Press, 2007.

Purcell, Jennifer E., Shin-ichi Uye, and Wen-Tseng Lo. "Anthropogenic Causes of Jellyfish Blooms and Their Direct Consequences for Humans: A Review." *Marine Ecology Progress Series* 350 (2007): 153.

Qin, Junjie, Ruiqiang Li, Jeroen Raes, Manimozhiyan Arumugam, Kristoffer Solvsten Burgdorf, Chaysavanh Manichanh, Trine Nielsen, et al. "A Human Gut Microbial Gene Catalogue Established by Metagenomic Sequencing." *Nature*, March 4, 2010, 59–65.

Quattro, Joseph M., William B. Driggers III, and James M. Grady. "*Sphyrna Gilberti* Sp. Nov., a New Hammerhead Shark (Carcharhiniformes, Sphyrnidae) from the Western Atlantic Ocean." *Zootaxa* 3702, no. 2 (2013): 159–78.

Rathbun, Richard. "Crustaceans, Worms, Radiates, and Sponges." In *The Fisheries and Fishery Industries of the United States*, section 1, *Natural History of Useful Aquatic Animals*, edited by George Brown Goode, 760–850. Washington, D.C.: Government Printing Office, 1884.

Reneerkens, Jeroen, Theunis Piersma, and Jaap S. Sinninghe Damsté. "Sandpipers (Scolopacidae) Switch from Monoester to Diester Preen Waxes during Courtship and Incubation, but Why?" *Proceedings of the Royal Society of London B* 269 (2002): 2135–39.

Richards, Eric J. "Inherited Epigenetic Variation—Revisiting Soft Inheritance." *Nature Reviews Genetics* 7, no. 5 (2006): 395–401.

Richardson, Sir John, William Swainson, and William Kirby. *Fauna Boreali-Americana; or, The Zoology of the Northern Parts of British America: The Birds.* London: J. Murray, 1831.

Riepe, Don. "An Ancient Wonder of New York and a Great Topic for Education." In *Limulus in the Limelight*, edited by John T. Tanacredi, 131–34. New York: Kluwer, 2001.

Rietschel, Ernst T., and Otto Westphal. "Endotoxin: Historical Perspective." In *Endotoxin in Health and Disease*, edited by Helmut Brade, Steven M. Opal, Stefanie N. Vogel, and David C. Morrison, 1–30. New York: Marcel Dekker, 1999.

Riley, John L. *Wetlands of the Ontario Hudson Bay Lowland.* Toronto: Nature Conservancy, 2011.

Rode, Karyn D., Eric V. Regehr, David C. Douglas, George Durner, Andrew E. Derocher, Gregory W. Thiemann, and Suzanne M. Budge. "Variation in the Response of an Arctic Top Predator Experiencing Habitat Loss: Feeding and Reproductive Ecology of Two Polar Bear Populations." *Global Change Biology*, January 1, 2014, 76–88.

Rode, Karyn D., James D. Reist, Elizabeth Peacock, and Ian Stirling. "Comments in Response to 'Estimating the Energetic Contribution of Polar Bear (*Ursus Maritimus*) Summer Diets to the Total Energy Budget' by Dyck and Kebreab (2009)." *Journal of Mammalogy* 91, no. 6 (2010): 1517–23.

Romero, Simon. "Peru Guards Its Guano as Demand Soars Again." NYTimes.com, May 30, 2008. http://www.nytimes.com/2008/05/30/world/americas/30peru.html?pagewanted=1&_r=0.

Ross, W. Gillies. "Whaling and the Decline of Native Populations." *Arctic Anthropology* 14, no. 2 (1977): 1–8.

Rossiter, Margaret W. *Women Scientists in America.* Baltimore: Johns Hopkins University Press, 1983.

Rudkin, David M. "The Life and Times of the Earliest Horseshoe Crabs." Presentation at the International Workshop on the Science and Conservation of Asian Horseshoe Crabs, Hong Kong, June 12, 2011.

Rudkin, David M., Graham A. Young, and Godfrey S. Nowlan. "The Oldest Horseshoe Crab: A New Xiphosurid from Late Ordovician Konservat-Lagerstatten Deposits, Manitoba, Canada." *Palaeontology* 51, no. 1 (2008): 1–9.

Rudloe, Jack. *The Wilderness Coast*. St. Petersburg, Fla.: Great Outdoors, 2004.

Runkle, Deborah. *Advocacy in Science: Summary of a Workshop Convened by the American Association for the Advancement of Science, Washington, DC, October 17–18, 2011*. Edited by Mark S. Frankel. American Association for the Advancement of Science, 2012.

Saey, Tina Hesman. "From Great Grandma to You: Epigenetic Changes Reach Down through the Generations." *Science News* 183 no. 7 (2013): 18–21.

Saffron, Inga. *Caviar*. New York: Broadway Books, 2002.

Saint-Exupéry, Antoine de. *Night Flight*. New York: Century, 1932.

Salemme, Mónica C., and Laura L. Miotti. "Archeological Hunter-Gatherer Landscapes since the Latest Pleistocene in Fuego-Patagonia." In *The Late Cenozoic of Patagonia and Tierra Del Fuego*, edited by J. Rabassa, 437–83. Amsterdam: Elsevier, 2008.

Sánchez, Marta I., Andy J. Green, and Eloy M. Castellanos. "Internal Transport of Seeds by Migratory Waders in the Odiel Marshes, South-west Spain: Consequences for Long-Distance Dispersal." *Journal of Avian Biology* 37, no. 3 (2006): 201–6.

Sanders, F., M. Spinks, and T. Magarian. "American Oystercatcher Winter Roosting and Foraging Ecology at Cape Romain, South Carolina." *Wader Study Group Bulletin* 120, no. 2 (2013): 128–33.

Schmidt, Niels M., Rolf A. Ims, Toke T. Høye, Olivier Gilg, Lars H. Hansen, Jannik Hansen, Magnus Lund, Eva Fuglei, Mads C. Forchhammer, and Benoit Sittler. "Response of an Arctic Predator Guild to Collapsing Lemming Cycles." *Proceedings of the Royal Society B: Biological Sciences*, November 7, 2012, 4417–22.

Schwarzer, Amy C., Jaime A. Collazo, Lawrence J. Niles, Janell M. Brush, Nancy J. Douglass, and H. Franklin Percival. "Annual Survival of Red Knots (*Calidris Canutus Rufa*) Wintering in Florida." *Auk*, October 1, 2012, 725–33.

"Sea Turtle Recovery Project." *Padre Island National Seashore*. http://www.nps.gov/pais/naturescience/strp.htm.

Seibert, Florence B. "Fever-Producing Substance Found in Some Distilled Waters." *American Journal of Physiology* 67, no. 1 (1923): 90–104.

———. *Pebbles on the Hill of a Scientist*. St. Petersburg, Fla.: [Florence B. Seibert], 1968.

Seino, Satoquo. "A Reconsideration of Horseshoe Crab Conservation Methodology in Japan over the Last 100 Years and Prospects for a Marine Protected Area Network in Asian Seas." Presentation at the International Workshop on the Science and Conservation of Asian Horseshoe Crabs, Hong Kong, June 12–16, 2011. In *Abstracts of Plenary Talks and Oral Presentations*, PT-3. Hong Kong, 2011.

Şekercioğlu, Çağan H. "Increasing Awareness of Avian Ecological Function." *Trends in Ecology and Evolution* 21, no. 8 (2006): 464–71.

Şekercioğlu, Çağan H., Gretchen C. Daily, and Paul R. Ehrlich. "Ecosystem Consequences of Bird Declines." *Proceedings of the National Academy of Sciences* 101, no. 52 (2004): 18042–47.

Seney, Erin E., and John A. Musick. "Historical Diet Analysis of Loggerhead Sea Turtles (*Caretta Caretta*) in Virginia." *Copeia* 2007, no. 2 (2007): 478–89.

Shin, Paul K. S., and Mark L. Botton from the IUCN Horseshoe Crab Species Specialist Group. Letter to the U.S. National Invasive Species Council, February 5, 2013.

Shuster, Carl N., Jr. "King Crab Fertilizer: A Once-Thriving Delaware Bay Industry." In *The American Horseshoe Crab*, edited by Carl N. Shuster Jr., Robert B. Barlow, and H. Jane Brockmann, 341–57. Cambridge, Mass.: Harvard University Press, 2003.

———. "A Pictorial Review of the Natural History and Ecology of the Horseshoe Crab *Limulus Polyphemus*, with Reference to Other Limulidae." *Progress in Clinical and Biological Research* 81 (1982): 1–52.

Shuster, Carl N., Jr., Mark L. Botton, and Robert E. Loveland. "Horseshoe Crab Conservation: A Coast-wide Management Plan." In *The American Horseshoe Crab*, edited by Carl N. Shuster Jr., Robert B. Barlow, and H. Jane Brockmann, 358–77. Cambridge, Mass.: Harvard University Press, 2003.

Sibley, David Allen. *The Sibley Field Guide to Birds of Eastern North America*. New York: Knopf, 2003.

Skagen, S. K., P. B. Sharpe, R. G. Waltermire, and M. B. Dillon. *Biogeographical Profiles of Shorebird Migration in Midcontinental North America: U.S. Geological Survey Biological Science Report 2000–0003*. Fort Collins, Colo.: U.S. Geological Survey, 1999.

Slocum, Joshua. *Sailing Alone around the World*, 1899. http://www.gutenberg.org/ebooks/6317.

Smallwood, K. Shawn. "Comparing Bird and Bat Fatality-Rate Estimates among North American Wind-Energy Projects." *Wildlife Society Bulletin* 37, no. 1 (2013): 19–33.

Smith, David R., Conor P. McGowan, Jonathan P. Daily, James D. Nichols, John A. Sweka, and James E. Lyons. "Evaluating a Multispecies Adaptive Management Framework: Must Uncertainty Impede Effective Decision-Making?" *Journal of Applied Ecology*, December 1, 2013, 1431–40.

Smith, David R., Michael J. Millard, and Ruth H. Carmichael. "Comparative Status and Assessment of *Limulus Polyphemus* with Emphasis on the New England and Delaware Bay Populations." In *Biology and Conservation of Horseshoe Crabs*, edited by John T. Tanacredi, Mark L. Botton, and David R. Smith, 361–86. Dordrecht: Springer, 2009.

Smith, Elizabeth H. "Colonial Waterbirds and Rookery Islands." In *The Laguna Madre of Texas and Tamaulipas*, edited by John W. Tunnell Jr. and Frank W. Judd, 183–97. College Station: Texas A&M University Press, 2002.

———. "Redheads and Other Wintering Waterfowl." In *The Laguna Madre of Texas and Tamaulipas*, edited by John W. Tunnell Jr. and Frank W. Judd, 169–81. College Station: Texas A&M University Press, 2002.

Smith, Fletcher M., Adam E. Duerr, Barton J. Paxton, and Bryan D. Watts. *An Investigation of Stopover Ecology of the Red Knot on the Virginia Barrier Islands*.

Center for Conservation Biology Technical Report Series, CCBTR–07–14. Williamsburg: College of William and Mary, 2008.

Smith, Hugh M. "Notes on the King-Crab Fishery of Delaware Bay." *Bulletin of the United States Fish Commission* (1989): 363–70.

Smith, Paul A., Kyle H. Elliott, Anthony J. Gaston, and H. Grant Gilchrist. "Has Early Ice Clearance Increased Predation on Breeding Birds by Polar Bears?" *Polar Biology*, August 1, 2010, 1149–53.

Specter, Michael. "Germs Are Us." *New Yorker*, October 22, 2012; 32–39.

Sperry, Charles. *Food Habits of a Group of Shorebirds: Woodcock, Snipe, Knot, and Dowitcher.* Wildlife Research Bulletin 1. Washington, D.C.: U.S. Department of the Interior, Bureau of Biological Survey, 1940. http://archive.org/details/foodhabitsofgrouoosper.

Sprunt, Alexander, Jr. "In Memoriam: Arthur Trezevant Wayne." *Auk* 48, no. 1 (1931): 1–16.

Sprunt, Alexander, Jr., and E. Burnham Chamberlain. *South Carolina Bird Life.* Columbia: University of South Carolina Press, 1949.

Stainsby, William. *The Oyster Industry of New Jersey.* Somerville: New Jersey Bureau of Industrial Statistics, 1902.

Stapleton, Seth, Elizabeth Peacock, David Garshelis, and Stephen Atkinson. *Aerial Survey Population Monitoring of Polar Bears in Foxe Basin.* Iqaluit: Nunavut Wildlife Research Management Board, 2012.

Stewart, D. B., and W. L. Lockhart. *An Overview of the Hudson Bay Marine Ecosystem.* Canadian Technical Report of Fisheries and Aquatic Sciences no. 2586, 2005.

Stirling, Ian, and Andrew E. Derocher. "Effects of Climate Warming on Polar Bears: A Review of the Evidence." *Global Change Biology* 18 (2012): 2694–2706.

Stokes, Donald, and Lillian Stokes. *Beginner's Guide to Shorebirds.* New York: Little, Brown, 2001.

Stone, Witmer. *Bird Studies at Old Cape May: An Ornithology of Coastal New Jersey.* Vol. 2. New York: Dover, 1965.

South Carolina Department of Natural Resources. "Horseshoe Crab Hand Harvest Permit HH14."

Subramanian, Meera. "An Ill Wind." *Nature*, June 21, 2012, 310–11.

Summers, R. W., L. G. Underhill, and M. Waltner. "The Dispersion of Red Knots *Calidris Canutus* in Africa—Is Southern Africa a Buffer for West Africa?" *African Journal of Marine Science* 33, no. 2 (2011): 203–8.

Sutton, Clay. "An Ecological Tragedy on Delaware Bay." *Living Bird* 22, no. 3 (2003): 31–37.

Sutton, George Miksch. "Birds of Southampton Island." *Memoirs of the Carnegie Museum* 12 (part 2, section 2) (1932): 1–275.

———. "The Exploration of Southampton Island, Hudson Bay." *Memoirs of the Carnegie Museum* 12 (part 1, section 1) (1932): 1.

Sutton, Scott V. W., and Radhakrishna Tirumalai. "Activities of the USP Microbiology and Sterility Assurance Expert Committee during the 2005–2010 Revision Cycle." *American Pharmaceutical Review* 14, no. 5 (2011): 12.

Swaddle, John P., and Stavros E. Calos. "Increased Avian Diversity Is Associated with Lower Incidence of Human West Nile Infection: Observation of the Dilution Effect." *PLoS ONE* 3, no. 6 (2008): e2488.

Swann, Benjie Lynn. "A Unique Medical Product (LAL) from the Horseshoe Crab and Monitoring the Delaware Bay Horseshoe Crab Population." In *Limulus in the Limelight*, edited by John T. Tanacredi, 53–62. New York: Kluwer, 2001.

Sweet, William, Chris Zervas, Stephen Gill, and Joseph Park. "Hurricane Sandy Inundation Probabilities Today and Tomorrow." *Bulletin of the American Meteorological Society* 94, no. 9 (n.d.): S17v–S20.

Székely, Csaba, Arjan Palstra, and Kalman Molnar. "Impact of the Swim-Bladder Parasite on the Health and Performance of European Eels." In *Spawning Migration of the European Eel*, edited by Guido van den Thillart, Sylvie Dufour, and J. Cliff Rankin, 201–26. New York: Springer Science and Business Media, 2009.

Szyf, Moshe. "Lamarck Revisited: Epigenetic Inheritance of Ancestral Odor Fear Conditioning." *Nature Neuroscience* 17, no. 1 (2014): 2–4.

"Table Supplies and Economics: What to Buy, When to Buy, and How to Buy Wisely and Well." *Good Housekeeping*, June 11, 1887.

Talmage, Stephanie C., and Christopher J. Gobler. "Effects of Past, Present, and Future Ocean Carbon Dioxide Concentrations on the Growth and Survival of Larval Shellfish." *Proceedings of the National Academy of Sciences* 107, no. 40 (2010): 17246–51.

Tebaldi, Claudia, Benjamin H. Strauss, and Chris E. Zervas. "Modelling Sea Level Rise Impacts on Storm Surges along US Coasts." *Environmental Research Letters*, March 1, 2012, 014032.

Thibault, Janet. *Assessing Status and Use of Red Knots in South Carolina: Project Report, October 2011–October 2013*. Charleston: South Carolina Department of Natural Resources, 2013.

Thibault, Janet, and Martin Levisen. *Red Knot Prey Availability: Project Report, March 2012–March 2013*. Charleston: South Carolina Department of Natural Resources, 2013.

Thomas, Lately. *Delmonico's: A Century of Splendor*. Boston: Houghton Mifflin, 1967.

Thomas, Lewis. *The Lives of a Cell*. New York: Penguin, 1978.

Thompson, Max. "Record of the Red Knot in Texas." *Wilson Bulletin* 70, no. 2 (1958): 197.

Thoreau, Henry David. *Cape Cod*. New York: Penguin, 1987.

Townsend, Charles Wendell. *Birds of Essex County*. Cambridge: Nuttall Ornithological Club, 1905.

Trull, Peter. "Shorebirds and Noodles." *American Birds*, June 1983.

Tsipoura, Nellie, and Joanna Burger. "Shorebird Diet during Spring Migration Stopover on Delaware Bay." *Condor* 101, no. 3 (1999): 635–44.

Tuck, James A. *Ancient People of Port au Choix: The Excavation of an Archaic Indian Cemetery in Newfoundland*. St. John's: Institute of Social and Economic Research, Memorial University of Newfoundland, 1976.

Tufford, Daniel L. *State of Knowledge: South Carolina Coastal Wetland Impoundments.* Charleston: South Carolina Sea Grant Consortium, 2005.

Tunnell, John W., Jr. "The Environment." In *The Laguna Madre of Texas and Tamaulipas,* edited by John W. Tunnell Jr. and Frank W. Judd, 73–84. College Station: Texas A&M University Press, 2002.

———— "Geography, Climate, and Hydrography." In *The Laguna Madre of Texas and Tamaulipas,* edited by John W. Tunnell Jr. and Frank W. Judd, 7–27. College Station: Texas A&M University Press, 2002.

Tuten, James H. *Lowcountry Time and Tide.* Columbia: University of South Carolina Press, 2010.

Ulrich, Glenn F., Christian M. Jones, W. B. Driggers, J. Marcus Drymon, D. Oakley, and C. Riley. "Habitat Utilization, Relative Abundance, and Seasonality of Sharks in the Estuarine and Nearshore Waters of South Carolina." *American Fisheries Society Symposium* 50, no. 125 (2007).

Urner, Charles A., and Robert W. Storer. "The Distribution and Abundance of Shorebirds on the North and Central New Jersey Coast, 1928–1938." *Auk* 66, no. 2 (1949): 177–94.

U.S. Centers for Disease Control and Prevention. "CDC Provides Estimate of Americans Diagnosed with Lyme Disease Each Year," August 19, 2013. http://www.cdc.gov/media/releases/2013/p0819-lyme-disease.html.

U.S. Department of Energy. *20% Wind Energy by 2030: Increasing Wind Energy's Contribution to U.S. Electricity Supply,* July 2008. http://www.nrel.gov/docs/fy08osti/41869.pdf.

U.S. Department of the Interior. "Secretary Jewell Announces $102 Million in Coastal Resilience Grants to Help Atlantic Communities Protect Themselves from Future Storms." Press release, June 16, 2014.

U.S. Fish and Wildlife Service. *Budget Justifications and Performance Information: Fiscal Year 2014.* U.S. Fish and Wildlife Service, Department of the Interior.

————. *Eskimo Curlew* (Numenius Borealis*). 5-Year Review: Summary and Evaluation.* Fairbanks: U.S. Fish and Wildlife Service, 2011.

————. *Piping Plover* (Charadrius Melodus*) 5-Year Review: Summary and Evaluation.* U.S. Fish and Wildlife Service Migratory Bird Publication R9–03/02. Arlington, Va., 2009.

————. "Proposed Threatened Status for the Rufa Red Knot (*Calidris Canutus Rufa*)." *Federal Register,* September 30, 2013, 60024–98.

————. "Rufa Red Knot Ecology and Abundance: Supplement to Endangered and Threatened Wildlife and Plants; Proposed Threatened Status for the Rufa Red Knot (*Calidris Canutus Rufa*)." Docket no. FWS-R5-ES-2013-0097; RIN 1018-AY17. *Federal Register,* September 30, 2013, 60023–60098.

————. "U.S. Fish and Wildlife to Restore Bay Beaches." USFWS Northeast Region press release, April 4, 2014.

U.S. Fish and Wildlife Service Shorebird Technical Committee. *Delaware Bay Shorebird–Horseshoe Crab Assessment Report and Peer Review.* U.S. Fish and Wildlife Service Migratory Bird Publication R9–03/02. Arlington, Va., 2003.

U.S. Food and Drug Administration. *Bad Bug Book: Foodborne Pathogenic Microorganisms and Natural Toxins*. 2nd ed., 2012.

————. *Guidance for Industry: Pyrogen and Endotoxins Testing: Questions and Answers*, June 2012.

Van Colen, Carl, Elisabeth Debusschere, Ulrike Braeckman, Dirk Van Gansbeke, and Magda Vincx. "The Early Life History of the Clam *Macoma Balthica* in a High CO2 World." *PLoS ONE* 7, no. 9 (2012): e44655.

Van Gils, Jan A., Phil F. Battley, Theunis Piersma, and Rudi Drent. "Reinterpretation of Gizzard Sizes of Red Knots World-wide Emphasises Overriding Importance of Prey Quality at Migratory Stopover Sites." *Proceedings of the Royal Society B: Biological Sciences*, December 22, 2005, 2609–18.

Van Roy, Peter, Patrick J. Orr, Joseph P. Botting, Lucy A. Muir, Jakob Vinther, Bertrand Lefebvre, Khadija el Hariri, and Derek E. G. Briggs. "Ordovician Faunas of Burgess Shale Type." *Nature* 465, no. 7295 (2010): 215–18.

Vaughan, Richard. *In Search of Arctic Birds*. London: T & A. D. Poyser, 1992.

Vézina, François, Tony D. Williams, Theunis Piersma, and R. I. Guy Morrison. "Phenotypic Compromises in a Long-Distance Migrant during the Transition from Migration to Reproduction in the High Arctic." *Functional Ecology* 26, no. 2 (2012): 500–512.

Vyn, Gerrit. "Spoon-Billed Sandpiper: Multimedia Resources." Cornell Lab of Ornithology (2011). http://www.birds.cornell.edu/Page.aspx?pid=2528.

Wakefield, Kirsten. *Saving the Horseshoe Crab: Designing a More Sustainable Bait for Regional Eel and Conch Fisheries*. Newark: Delaware Sea Grant, 2013.

Waldbusser, George G., Elizabeth L. Brunner, Brian A. Haley, Burke Hales, Christopher J. Langdon, and Frederick G. Prahl. "A Developmental and Energetic Basis Linking Larval Oyster Shell Formation to Acidification Sensitivity." *Geophysical Research Letters* 40, no. 10 (2013): 2171–76.

Waldbusser, George G., and Joseph E. Salisbury. "Ocean Acidification in the Coastal Zone from an Organism's Perspective: Multiple System Parameters, Frequency Domains, and Habitats." *Annual Review of Marine Science* 6 (2014): 221–47.

Walsh, J. J., C. R. Tomas, K. A. Steidinger, J. M. Lenes, F. R. Chen, R. H. Weisberg, L. Zheng, J. H. Landsberg, G. A. Vargo, and C. A. Heil. "Imprudent Fishing Harvests and Consequent Trophic Cascades on the West Florida Shelf over the Last Half Century: A Harbinger of Increased Human Deaths from Paralytic Shellfish Poisoning along the Southeastern United States, in Response to Oligotrophication?" *Continental Shelf Research* 31, no. 9 (2011): 891–911.

Wang, Zhaohui Aleck, Rik Wanninkhof, Wei-Jun Cai, Robert H. Byrne, Hu Xinping, Tsung-Hung Peng, and Wei-Jen Huang. "The Marine Inorganic Carbon System along the Gulf of Mexico and Atlantic Coasts of the United States: Insights from a Transregional Coastal Carbon Study." *Limnology and Oceanography* 58, no. 1 (2013): 325–42.

Watts, Bryan D. *Wind and Waterbirds: Establishing Sustainable Mortality Limits within the Atlantic Flyway*. Center for Conservation Biology Technical Report Series,

CCBTR–05–10. Williamsburg: College of William and Mary / Virginia Commonwealth University, 2010.

Watts, B. D., F. M. Smith, T. Keyes, E. K. Mojica, J. Rausch, B. Truitt, and B. Winn. "Whimbrel Tracking in the Americas." wildlifetracking.org/whimbrels.

Watts, Bryan D., and Barry R. Truitt. "Decline of Whimbrels within a Mid-Atlantic Staging Area (1994–2009)." *Waterbirds* 34, no. 3 (2011): 347–51.

Wayne, Arthur Trezevant. *Birds of South Carolina.* Charleston: Charleston Museum, 1910.

Weber, Louise M., and Susan M. Haig. "Shorebird Use of South Carolina Managed and Natural Coastal Wetlands." *Journal of Wildlife Management* 60, no. 1 (1996): 73.

Wellnhofer, Peter. *Archaeopteryx: The Icon of Evolution.* Translated by Frank Haase. Munich: F. Pfeil, 2009.

Wenny, Daniel G., Travis L. Devault, Matthew D. Johnson, Dave Kelly, Çağan H. Şekercioğlu, Diana F. Tomback, and Christopher J. Whelan. "The Need to Quantify Ecosystem Services Provided by Birds." *Auk* 128, no. 1 (2011): 1–14.

West-Eberhard, Mary Jane. "Developmental Plasticity and the Origin of Species Differences." *Proceedings of the National Academy of Sciences of the United States of America* 102, supplement 1 (2005): 6543–49.

Wetlands International. "*Calidris Canutus.*" *Waterbird Population Estimates,* 2013. wpe.wetlands.org.

Wetmore, Alexander. *Our Migrant Shorebirds in Southern South America.* U.S. Dept. of Agriculture Technical Bulletin 26. Washington, D.C.: U.S. Government Printing Office, 1927.

Whelan, Christopher J., Daniel G. Wenny, and Robert J. Marquis. "Ecosystem Services Provided by Birds." *Annals of the New York Academy of Sciences* 1134, no. 1 (2008): 25–60.

Wheye, Darryl, and Donald Kennedy. *Humans, Nature, and Birds.* New Haven: Yale University Press, 2008.

WHSRN. "Chaplin Old Wives Reed Lakes," 2009. http://www.whsrn.org/site-profile/chaplin-old-wives-reed-lakes.

Wildfowl and Wetlands Trust. "Saving the Spoon-Billed Sandpiper," 2014. http://www.saving-spoon-billed-sandpiper.com/.

Williams, Glyndwr, ed. *Andrew Graham's Observations on Hudson's Bay, 1767–91.* London: Hudson's Bay Record Society, 1969.

Williams, S. Jeffress, Kurt Dodd, and Kathleen Gohn. "Coasts in Crisis." U.S. Geological Survey Circular, 1997. http://pubs.usgs.gov/circ/c1075/hog.html.

Wilson, Alexander. *The Life and Letters of Alexander Wilson.* Edited by Clark Hunter. Memoirs, vol. 154. Philadelphia: American Philosophical Society, 1983.

Wilson, Alexander. *Wilson's American Ornithology: With Notes by Jardine; to Which Is Added a Synopsis of American Birds, Including Those Described by Bonaparte, Audubon, Nuttall, and Richardson.* Edited by T. M. Brewer. Boston: Otis, Broaders, 1840.

Wilson, E. G., K. L. Miller, D. Allison, and M. Magliocca. *Why Healthy Oceans Need Sea Turtles.* Oceana, 2010.

Wilson, N. C., and D. McRae. *Seasonal and Geographical Distribution of Birds for Selected Sites in Ontario's Hudson Bay Lowland.* Toronto: Ontario Ministry of Natural Resources, 1993.

Woodin, Marc C., and Thomas C. Michot. "Redhead (*Aythya Americana*)." Edited by A. Poole and F. Gill. *The Birds of North America Online,* 2002.

Yang, Hong-Yan, Bing Chen, Mark Barter, Theunis Piersma, Chun-Fa Zhou, Feng-Shan Li, and Zheng-Wang Zhang. "Impacts of Tidal Land Reclamation in Bohai Bay, China: Ongoing Losses of Critical Yellow Sea Waterbird Staging and Wintering Sites." *Bird Conservation International* 21, no. 3 (2011): 241–59.

Young, Graham A., David M. Rudkin, Edward P. Dobrzanski, Sean P. Robson, and Godfrey S. Nowlan. "Exceptionally Preserved Late Ordovician Biotas from Manitoba, Canada." *Geology* 35, no. 10 (2007): 883–86.

Zhang, Xinzhi, Martin I. Meltzer, César A. Peña, Annette B. Hopkins, Lane Wroth, and Alan D. Fix. "Economic Impact of Lyme Disease." *Emerging Infectious Diseases* 12, no. 4 (2006): 653–60.

Zimmer, Carl. "The Price Tag on Nature's Defenses." *New York Times,* June 10, 2014, D3.

Zimmer, Kevin J. *A Birder's Guide to North Dakota.* Denver: L & P, 1979.

Zobel, R. D. "Memorandum of Decision: *Associates of Cape Cod, Inc. and Jay Harrington v. Bruce Babbitt.*" U.S. District Court, District of Massachusetts, May 22, 2001.

Zöckler, Christoph, Tony Htin Hla, Nigel Clark, Evgeny Syroechkovskiy, Nicolay Yakushev, Suchart Daengphayon, and Rob Robinson. "Hunting in Myanmar Is Probably the Main Cause of the Decline of the Spoon-Billed Sandpiper *Calidris Pygmeus.*" *Wader Study Group Bulletin* 117, no. 1 (2010): 1–8.

Zöckler, C., R. Lanctot, and E. Syroechkovsky. "Waders (Shorebirds)." *Arctic Report Card: Update for 2012,* February 2013. www.arctic. noaa.gov/reportcard/waders. html.

Acknowledgments

The Canadian Wildlife Service and the National Wildlife Research Centre of Environment Canada, the Curtis and Edith Munson Foundation, the Norcross Wildlife Foundation, the Ocean Foundation, and the Wellesley College Elvira Stevens Travelling Fellowship enabled me to follow red knots, making the book possible. So many of you graciously and unstintingly shared your hospitality, your time, and your wisdom. Your ideas, research, and commitment are an inspiration and the backbone of this book. I have tried to reflect your work accurately and fairly: mistakes that remain are mine.

Charles Duncan from the Manomet Center for Conservation Sciences introduced me to red knot researchers everywhere. Brian Harrington, author of the lovely *Flight of the Red Knot*, Guy Morrison, and Larry Niles generously shared their work along the length of the flyway. For the knots' winter home in Tierra del Fuego, I was grateful for the time and perspective of Carmen Espoz Larrain, Boris Cvitanic, the men from Empresa Nacional del Petróleo (ENAP) in Cerro Sombrero and Posesión, Ricardo Matus, Ricardo Olea, Diego Luna Quevedo, Roy Hann, Edward Owens. In Río Gallegos, Argentina: Silvia Ferrari, Carlos Albrieu. In Las Grutas and San Antonio Oeste, Argentina: Patricia M. González, Mirta Carbajal; Liz Assef, Silvana Sawicki; Anabel Chávez, Yanina Lillo, Gabriela Mansilla, Guadalupe

Sarti, Luciana Ceccacci Sawicki; Gimena Mora, Amira Mandado; Anahí
Valverde, Horacio García, Inalafquen, Vuelo Latitud 40. From Eco Huellas
in General Roca, Argentina: Cande Lorente, Maria Belén Pérez, Emi Suarez.

In Texas: David Newstead, Tony Amos, Billy Sandifer, Anse Windham,
Paul Zimba; Wes Tunnell, Kim Withers; Jim Blackburn, Ron Outen; Kelly
Fuller, Shawn Smallwood; Donna Shaver; Ruth Kelley. Along the central
flyway and prairies: Cheri Gratto-Trevor, Scott Wilson; Doug Backlund, Joe
Gryzbowski, Lawrence Igl, Dan Svingen, Jeff Palmer, Max Thomson. In
Florida: Doris and Pat Leary; Ron Smith; Lynne Knauf, Bob Greenbaum. In
Georgia: Tim Keyes, Brad Win; Wendy Paulson, Stacia Hendricks, Bonnie
Hilton, Abby Sterling, the team at Little St. Simons. In South Carolina: Felicia
Sanders, Al Segars; Sarah Dawsey, Nathan Dias, William Driggers, Aaron
Given, Dean Harrigal, Jim Jordan, Craig LeSchack, Jamie Rader, Pete Richards,
Michael Slattery; Ernie Wiggers, Ellen Solomon and Richard Wyndham.

For horseshoe crabs: Jim Cooper, John Dubczak, Barbara Edwards,
Jerry Gault, Jill Schultz, Daniel Yokell; Jeanne Boylan, Larry Delancey, Brad
Floyd; Mark Botton, Glenn Gauvry, John Tanacredi; Eric Hallerman; Allen
Burgenson, Maribeth Janke; Jeak Ling Ding; Conor P. McGowan, David
Smith; William McCormick, Robert Mello, Radhakrishna Tirumalai; Karen
Zink McCullough; Tom Novitsky. In Virginia: Fletcher Smith, Barry Truitt,
Bryan Watts. In Delaware and New Jersey: Kevin Kalasz, Richard Weber,
Nigel Clark; Amanda Dey, David Mizrahi, Susan Kraham, David Wheeler;
David Stallknecht, Angela Maxted, Scott Kraus, Pejman Rohani; William
Sweet; Mike Haramis; Suzann and John Callinan, Barry Camp, Willets
Corson Camp, Frances Camp Hansen, Pete Dunne, Betsy Haskin, Marjory
Nelson, John and Lorraine Nicholas, Pat and Clay Sutton; Sandra Axelsson,
Jamie Hand, Carole Mattessich Raritz, Donna Soffee, Kristoffer Whitney. In
Massachusetts: Kate Iaquinto, Stephanie Koch, Robin Lepore, Bud Oliveira;
Bob Prescott; Mary-Jane James-Pirri; Colleen Coogan; Kathryn Heinze.

For horseshoe crabs and knots through time and the geology of
Hudson Bay: Deborah Buehler, David Corrigan, Rob Fensome, Andy Fyon,
David Rudkin, Graham Young, Peter van Roy. For epigentics and adapta-
tion: Theunis Piersma, Michael Skinner. For why we need birds: Andy
Green, Doug McCauley.

In the Arctic: Grant Gilchrist, Paul Smith, and the National Wildlife
Research Centre of Environment Canada; Alannah Kataluk-Primeau,

Naomi Man in 't Veld, Meagan McCloskey, Josiah Nakoolak, Kara Anne Ward; Amie Black, Frankie Jean-Gagnon, Holly Hennin, Mike Jannssen, and the eider camp; Ken Abraham, Jim Leafloor; Tony Gaston, Kyle Elliot, Sam Iverson, Karyn Rode; Darryl Edwards, Siu-Ling Han, Jennie Rausch.

On the route home in James Bay: Christian Friis and the Canadian Wildlife Service, Mark Peck; Jennifer Goulet, Mike and Ken Burrell; Jean Iron, Barbara Charlton, Andrew Keaveney, Ian Sturdee, Josh Vandermeulen. In Mingan: Yves Aubry, Steve Gates, Pierrot Vaillancourt, Amélie Robillard, Ilya Klvana. In Suriname: Arie Spaans. For shorebird and knot population trends, and for spoon-billed sandpipers: Brad Andres. For godwits, Robert Gill. For data loggers, the new radio tags, satellite trackers: Ron Porter and Matthew Danihel, Phil Taylor and Ann McKellar, Paul Howey.

For insight and clarity framing the broader issues: Don Kennedy. MIT continues to support my work in many ways. An appointment as visiting scholar there enables me to undertake the extensive research required for this kind of book. Mary Sears at the Ernst Mayr Library at Harvard's Museum of Comparative Zoology and Fred Burchsted at Widener led me to sources I would not have found otherwise. Maria Silvia Rodrigo assisted early on with translation, and later Barbara Kelley, at all hours of the day and night, provided more hours of translation than she or I ever imagined would be needed, in letters, in transcontinental and transatlantic skype calls, and in Argentina, where I would have been lost without her. Thank you to Sylvan LaChance and Wendy Quinones for generously copyediting.

I am grateful for the support and enthusiasm of agents Wendy Strothman and Lauren McLeod, of Bill Nelson and Michael DiGiorgio, of John Marzluff, and of senior executive editor Jean Thomson Black and the team at Yale University Press, including Susan Laity, Nancy Ovedovitz, and Robin DuBlanc, who took such care shepherding this book through publication, and to Liz Pelton, for bringing it out into the world.

Wayne Petersen taught me about shorebirds of Cape Cod, and Chris Leahy was always on call to answer questions avian. Weekend after weekend and with kindness and endless patience, Robert Buchsbaum taught me how to find birds on Plum Island and in Gloucester. Derek Brown and Prita M. Manganiello did the same in the bay. Diana Peck introduced me to Essex Bay so many years ago, providing the first boat. Since then, every season,

Don Parsons generously shares his dock. Wendy Williams and Margaret Quinn, your support is invaluable. To Susan Troyan, Elsie Levin, and Hal Burstein, thank you.

To Abby and Susannah, and to Dan, thank you for precise, careful reading and sharp insight and, when I didn't think I could write one more word, for carrying this book, and me, home. Dan, you are the best partner, always. In the light of your unwavering optimism, anything is possible.

Index

Note: An "f" following a page number indicates a figure.